GOLD IN BRITAIN

by

R M CALLENDER

GOLDSPEAR (UK) LIMITED

BEACONSFIELD

First published 1990
A Goldspear (UK) Ltd Publication

This book is copyright under the Berne Convention. All rights reserved. Apart from any fair dealing for the purpose of private study, research, criticism or review – as permitted under *The Copyright Act 1989*, no part of this publication may be reproduced, stored in a retrieval system or transmitted in any form or by any means, electronic, electrical, chemical, mechanical, optical photocopying, recording or otherwise, without the prior permission of the copyright owners. Enquiries should be addressed to
Goldspear (UK) Ltd,
Box 203, Beaconsfield,
Buckinghamshire HP9 2TQ, England

© Ronald Montgomery Callender 1990
ISBN 0 9514134 2 2

DEDICATION

To my mother, Isa, who has encouraged, supported and followed my ideas, plans and activities for over half a century.

Typeset by Riverhead Typesetters, Grimsby, South Humberside.
Printed and bound by Hollen Street Press Limited, Slough, Berkshire.

MADE AND PRINTED IN GREAT BRITAIN

CONTENTS

	Page
Foreword	vi
Acknowledgements	vii
Introduction	ix
The Yellow Metal	ix
The Fourth Degree	xii
Prospecting by Computer	xv

PART I
ENGLAND AND WALES

England's Gold	2
Gold of South Wales	5
The Dolgellau Gold Belt of North Wales	10

PART II
SCOTLAND

Scotland's Gold	20
Crawfordmoor	20
The Golden Zone	28
The Kildonan Gold Rush	30
After the Gold Rush	36
Letter from Inverness	39
Gold from Kildonan	44
Nuggets	47

PART III
APPENDIX

Panning	52
The Henderson Pump	55
Prospecting with a Sluice Box	56
The Electronic Prospector	58
Rivers and River Lore	60
All is not Gold	61
The Future	63

LIST OF ILLUSTRATIONS

	Page
1. Share certificate of Clogau Gold Mines PLC.	xi
2. A 2.5 gm bar of gold produced by Johnson Mathey.	xii
3. Gold nuggets recovered from a Scottish river.	xii
4. Gold locations along the 4° West of longitude.	xiii
5. Image of the distribution of Arsenic over the Argyll region *(extended caption on page v).*	xv
6. A rare specimen of gold from Hope's Nose, Devon.	2
7. Interior of Goldscope Mine, Newlands Valley, Cumbria.	3
8. Roman adit at the Ogofau Gold Mine, Dolaucothi, South Wales.	6
9. Activity at the Dolaucothi mine, *Circa* 1933.	6
10. Entrance to Mitchell's adit, Dolaucothi.	7
11. The Turf Mine at Dolfrwynog.	11
12. St David's vein of the Clogau Gold Mine.	12
13. Cover of Certificate of Authenticity, for "Welsh Maiden" products.	14
14. Centre pages of Certificate of Authenticity.	14
15. The "Welsh Maiden" trade mark on a billet of gold.	14
16. Group of 19th Century gold miners.	15
17. Tyn-y-Cornel level at Clogau Gold Mine.	15
18. Main adit of the Cefn Coch Gold Mine.	16
19. Interior view of the Gwynfynedd Gold Mine.	16
20. Forlorn symbol of the Gwynfynedd Gold Mine.	17
21. 1635 map of Crawfordmoor by John Blaeu.	20
22. Drawing of the Scottish Crown by the author.	21
23. Scottish crown used on an insurance plaque.	21
24. Example of gold ducat featuring King James V in "capberet".	21
25. Reproduction of first folio of the Cottonian manuscript.	23
26. Keeper ring made from Scottish gold.	24
27. Princess Mary of Teck displaying Scottish gold ring.	24
28. John Blackwood who discovered the largest nugget found in Scotland for 200 years.	25
29. "Prospectors" pose for a snapshot *circa* 1928.	26
30. A modern day prospector.	26
31. Carved stone doorway slab in George Heriot School, Edinburgh.	27
32. Disclosure of the "Golden Zone", 1988.	28
33. Press cuttings.	29
34. Panning in the Cononish River.	29
35. Portrait of Robert Nelson Gilchrist.	30
36. Announcement of the discovery of gold in Sutherland.	31
37. Facsimile of licence to mine for gold for one month.	32
38. The "town" of Baile an Or.	33

39. Duke of Sutherland presenting gold watch to Mr Robert Gilchrist. 33
40. An account of Sutherlandshire Gold Diggings. 34
41. Advertisement in *The Inverness Courier* July 1869. 35
42. Advertisement signalling the end of the Gold Rush. 35
43. Downstream view of the Kildonan Burn. 36
44. Three men operating a sluice box. 37
45. Owner of Suisgill Estate supervising sluicing operation in 1914. 38
46. Artist's impression of interior of Kildonan Parish Church. 40
47. 1869 sketch of "Rocking the Cradle". 41
48. Interior of the store at Kildonan. 42
49. Pendant made from Kildonan gold. 45
50. The Bishop's cross. 46
51. The Rutherford Nugget. 48
52. The Helmsdale Nugget. 48
53. The Sutherland Nugget. 48
54. 1875 pamphlet cover showing the Gemmel Nugget. 49
55. A comfortable "seat" when panning. 52
56. Chris Engels of Aberdeen – photograph taken in 1983. 53
57. Patrick Reason of North Wales panning in 1987. 54

Extended caption to illustration 5 on page xv
This image is derived from the determination of stream sediment samples collected by the Geochemical Survey Programme and converted into image format. An image database which can be interrogated in real time is available through the GISA mineral exploration service of the British Geological Survey. This image is subject to NERC copyright. Geochemical atlases of this data are also available. *(Illustration reproduced by courtesy of British Geological Survey.)*

v

FOREWORD

To many people, the idea of finding gold in Britain is one of disbelief and was first encountered by John Calvert:

> In the more modern gold discoveries . . . the community at large did not acknowledge the fact till some light burst upon them . . . it is almost in vain to tell them there is gold or anything else in the country, if they are disposed to believe there is not.
>
> The transmutation of gold they may be willing to receive, but the existence of gold bodily they deny. Gold is a rare metal, gold is only to be found in certain countries, gold is only to be found in hot countries — gold never was found in Britain and never will be — if it were Peru, that is a different thing altogether.
>
> Show them a specimen — they do not admit the presence of gold unless they cannot help it. It is pyrites, or it may be anything but it cannot be gold. If they allow there is gold, there cannot be much of it, and it will never pay.

In spite of the attitude, (which still exists today), when a gold symposium was held at Salford University in May 1987, ninety persons from all over Britain attended for a day of lectures, demonstrations and displays of equipment and photographs. Two years later, an equal number participated in an open air weekend by the banks of the rivers Mawdach and Afon Wen where the emphasis was on gold panning, tuition and gold panning competitions.

It is to those persons that this book is addressed.

ACKNOWLEDGEMENTS

For some reason there has been no book on the subject of gold in Great Britain since 1853 when an Australian mining engineer, John Calvert published *The Gold Rocks of Great Britain and Ireland and a General Outline of the Gold Regions of the World with a Treatise on the Geology of Gold**. From time to time we hear rumours of books in preparation but they fail to materialise. Yet, when George W Hall wrote *The Gold Mines of Merioneth* in 1975, it was an immediate success and has now been issued as a second edition with equal success.

This venture is an experiment in achieving a gold prospector's handbook. In the days of the great gold rushes of the 19th century, many publishers responded with manuals on do-it-yourself gold prospecting. Many were, of course, worthless but filled a need at the time and yet, nowadays, are eagerly-sought collector's items. In making a start on this handbook and seeing the task to completion, I have been helped and hindered by many people in the world's gold communities. John Stephens of Eastbourne was instrumental in planting the germ of the idea while we attended the eleventh World Goldpanning Championships in Foix in the south of France, and as we discussed the proposition, further encouragement came from Staffan Feron of the Goldpanners' Association, Beaconsfield, England.

My colleagues at Wanlockhead Museum Trust, Geoff Downs-Rose of Wanlockhead and Bill Harvey of Leadhills, have both provided much encouragement to my gold interests and each has extended the scope of my activities by providing — as necessary — finance, bed, breakfast, information and opportunity for serious discussion.

For a long time Rex Bingham of Macclesfield has spoken of gold prospecting as a hobby and has generously contributed ideas, knowledge and experience. Alf Henderson of Cumbria is an enthusiast who recovers much gold from the British rivers, and sets a good example in fostering the interchange of ideas and the friendship which comes from recreational gold prospecting. It has to be said, however, that he works hard for his rewards and has already bequeathed his gold collection to the British Museum.

*Now available in a facsimile reprint volume from Goldspear (UK) Ltd.

Many other people have contributed in different ways to my comprehension and understanding of the gold scene. In Finnish Lapland, Kauko Launonen, manager of the Gold Village, and Inkeri Syrjanen, curator of the Gold Museum at Tankavaara, have established a strong base for organised gold panning in Europe. Walking in the wilderness of Finnish Lapland with Reima Paakkanen, author of the *Finnish Gold Prospector's Handbook*, established the perfect conditions for discussion, more discussion and even more discussion on our mutual interests.

The enthusiasm of the Swiss gold prospectors — Victor Jans, Peter Pfander, Rudi Steiner, Roland Brunner, — demonstrates that an interest in gold does not end at panning but can extend to history, to geology, to collecting, to travel, to photography, to friendship and to good food.

The doyen of all British prospectors, octogenarian Percy Collins, has maintained his passion for gold panning for over fifty years and yet continues to impress and motivate today's enthusiasts. His personal contribution to recreational gold mining should never be underestimated.

INTRODUCTION

THE YELLOW METAL

The chemical symbol for gold is 'Au' and is derived from the Latin word 'aurum' which comes from Aurora, the goddess of the shining dawn.

Throughout the ages, gold has been prized because of its beauty, permanence and rarity. It has served as an indicator of wealth, as money for the buying and selling of merchandise, and has been much used for jewellery and other items of decoration.

Gold has special properties which makes it useful in many fields of human endeavour. It is resistant to corrosion, it is the most malleable and ductile of all metals, and it is a good conductor of heat and electricity. Gold can be bonded easily to other metals and it will mix readily with other metals to form alloys which, in turn, can possess exceptional properties.

In the twentieth century, gold has an important place in industry, aerospace research, dentistry and medicine, electronics and packaging.

Gold is so soft that one ounce can be stretched into a wire measuring fifty miles long, or be hammered into a sheet so thin that it will cover 100 square feet. It is so rare that estimates suggest that no more than 88,000 tons have been taken from the earth throughout history and that this amount could be contained in a cube with eighteen yard sides. More steel is poured in one hour than gold has been poured since the beginning of time. Since it does not rust, tarnish or corrode, gold virtually lasts for ever.

The Celtic tribes of the sixth and seventh centuries who inhabited Southern Scotland possessed large quantities of gold and the contemporary bards relate how the gold torque was standard wear for all the principal men. Two collar-like ornaments, made of gold and found near the borders of Coulter Parish, are displayed in the National Museum of Scotland, Edinburgh.

The first-ever coins originated in the sixth century B.C. when King Croesus (one of the rulers of the Persian Empire) gave instructions that his personal heraldic device should be stamped on all the gold bars in his possession.

This 'coin' became known as the Stater and from it developed the concept that a coin could be the basis of economic life. In Imperial

Rome gold coins were the cornerstone of much of the trading which developed throughout the far-flung empire. The most common denomination was known as the Aurus and it was a tradition that each Emperor should have at least one design of coin issued throughout his reign.

In 1489, when Henry VII commissioned a coin which would convey the majesty and dignity of the English throne and the spirit and pride of his Tudor dynasty, gold was selected as the metal and the new coin took the name 'sovereign' from the splendid obverse design which showed the King in all his majestic glory, crowned, enthroned and in full coronation regalia. The tradition of the sovereign continues to this day and gold continues to be used by the Royal Mint.

In order to cater for a demand for bullion, the Government introduced legislation in 1987 which permitted the issue of a new gold coin — the Britannia. There are four Britannia coins available, each of which is legal tender by Royal Proclamation, with face values of £100, £50, £25 and £10. Their fine gold content is respectively 1 ounce, 1/2 ounce, 1/4 ounce and 1/10 ounce.

The term 'karat' is used to designate the proportion of fine gold in an alloy. The word karat, and carat, derives from the Italian 'carato', the Arabic 'qirat', and the Greek 'keration', all of which meanings describe the fruit of the carob tree. The seeds of the fruit were once used to balance the scales used for weighing gems and gold in oriental bazaars.

In the United States, the proportion of gold in an alloy is expressed in karats — often abbreviated to 'k' — using a scale of 24 so that 24k is 100% gold. Therefore, 18k indicates 18 parts of gold and 6 parts of other metals. In the European systems, however, fineness of gold is expressed on a scale of 1000 so that, for example 24k is 100% pure and has a fineness of 999. 18k gold is 75% pure and has a fineness of 750, and 14k gold is 58.5% pure with a fineness of 585. In Great Britain, the legal minimum is 9k gold.

Gold has the necessary requirement of currency and, for many years, governments backed their issues with gold. This became known as 'the gold standard' but the disruptions of the 1930's forced the United States to make it illegal for private persons or companies to own gold bullion. In 1970, gold backing for the U.S. dollar had been reduced to 25% of value and the standard was finally removed. The following year, President Gerald Ford allowed U.S. citizens to deal again in gold.

The price of gold is determined in London. Twice a day, representatives of the London Gold Market meet at Rothschild Bank and agree the price as a result of the buying and selling orders which are held by the member firms. When supply and demand are balanced, the price is said to be 'fixed' and the information is then relayed by wire services around the world.

There are three accepted ways to invest in gold:

* Ownership of gold bars and wafers which range in size from 400 ounce 'good delivery bars' to 1/4 ounce wafers. Each bar is identifiable and is always at least 99.5% pure gold.

* Investment in gold coins, which may either be numismatic or bullion. The latter are popular investments because they can be sold more easily than the gold bar.

* Purchase of shares in a gold mining company. A successful gold mine, however, is a decreasing asset.

Fool's gold is a popular name for iron pyrites and can be mistaken for gold as it is often associated with it. Iron pyrites are hard and brittle and possess a more crystalline appearance than gold.

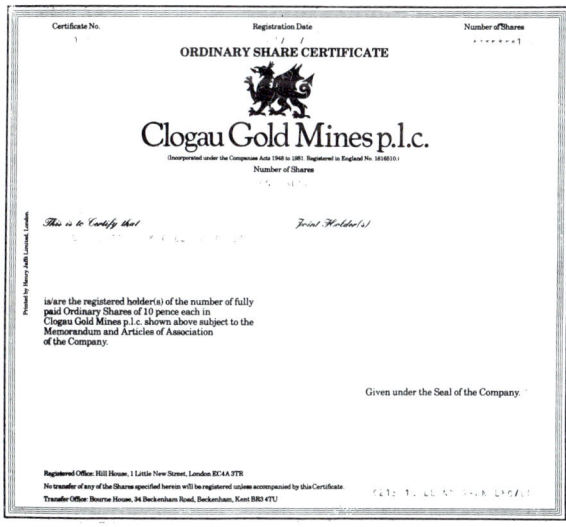

1. A share certificate of Clogau Gold Mines plc issued when the company joined the Unlisted Securities Market in 1984.

2. (*Left*) A bar of gold weighing 2.5 gm produced for sale by Johnson Mathey which carries its registration mark. 3. (*Right*) Gold nuggets recovered from a Scottish river by Mr G A Henderson of Cumbria.

THE FOURTH DEGREE

In spite of regular accounts which associate the discovery of gold in Great Britain with the line of longitude, 4° West, there are seldom explanations for this coincidence. A skilled geologist can provide a plausible explanation based on the theory of plate tectonics by assuming that the formation of the Caledonides, or Caledonian Mountains when North America collided with Europe 400 million years ago, created the ideal conditions. Magma, from deep below the surface of the earth, would be forced into cracks and fissures where it would solidify in conjunction with folded rocks to become mountainous terrain capped by volcanoes.

During the cooling action, the magma and its adjacent rocks would evolve as gold and, in time, the scouring action of wind and rain would expose the gold and wash it into the rivers where it is found nowadays.

In October 1987, a theory developed by Dr T O'Driscoll, a consultant to the Australian Western Mining Corporation, has attempted to explain the distribution of mineral deposits by recognising linear patterns on the earth's crust [1]. He claims that many of the world's ore deposits are located on such lineaments and he gives examples for uranium, nickel and copper. Although the theory is welcomed cautiously by other geologists as being "radical and innovative", it does establish a regular grouping of terrestrial lineaments which relate to the occurrence of ores.

4. A map of Britain which demonstrates the alignment of the principal gold locations in the proximity of the 4° West of longitude.

The Australian Academy of Technical Sciences elected Dr O'Driscoll to Fellowship in 1984 for the successful application of his ideas and in the absence of other evidence, it does help to justify the alignment of gold along the fourth degree West of longitude in Britain.

However, it is difficult to reconcile any such theory to the gold which is found at Kildonan in the north of Scotland. The source has never been found and the differing explanations which followed on its discovery in 1868 have never been brought up to date. There are two schools of thought; each is backed by distinguished and experienced men.

William Cameron [2] and Rev J M Joass [3] supported the proposition that the gold had a "local origin":

> We have in Kildonan and Suisgill chloritic micaceous flags, gneiss, quartzite, and granitic rocks. The drift found in the washdirt has no speciality pervading it which points to any other than

xiii

a local origin; indeed, it seems strictly to correspond with the rocks in the neighbourhood [2].

A cautious Sir Roderick Murchison [4] and a reckless John Campbell [5] believed, with different degrees of enthusiasm, that the gold had been imported by glacier action. Sir Roderick's thesis was:

By means of glacial action, the great mass of detritus from the Western and Central Highlands has been transported eastward; the gold debris found near Kildonan is the result of abrasion of granitic and metamorphic Lower Silurian rocks in the interior.

John Campbell, in his eminently readable account of the gold fields, states with conviction:

Ice moved from the northern end of Scandinavia south-westwards to Scotland, south-eastwards through Finland. But the strange coincidence in this gold chapter in the history of northern drift is, that the river Tana in Russian Lappland yields gold, and was the scene of busy digging and unsuccessful prospecting for quartz veins in the autumn of 1868. According to this larger view, Sutherland gold may have come from Lappland if it belongs to the northern drift.

The reason for including these extracts from a 19th century argument is not to resolve the controversy but to question our 20th century acceptance of the origin of Kildonan's gold.

William Cameron deserves the last word:

Upon the whole, the question is very perplexing, and, so long as gold in situ remains undiscovered it will always remain an open one. Glasgow, 27th June 1870

References
1. Patterns in the Crust : a Key to Ore Discovery
Jonathan Selby   *Geology Today*   September/October 1987
2. On the Sutherlandshire Gold Fields
Wm Cameron   Glasgow Geological Society 1870
3. Notes on the Sutherland Goldfield
J M Joass (with an introduction by Sir R R Murchison)
Quarterly Journal of the Geological Society 1869
4. Siluria Fourth Edition 1867 pp 448 — 472
5. Something from "The Diggins" in Sutherland
John Campbell of Islay, Author of "Frost and Fire" from Odds and Ends, Series No. 22 1869

PROSPECTING BY COMPUTER

In February 1989, Martin Belderson[1], a freelance writer specialising in geology, disclosed a modern method of "computer prospecting" which had its origins in a scheme initiated by the British Geological Survey of Great Britain whilst searching for uranium in the north of Scotland in the late 1960s. Through employing a technique based on the analysis of the geochemical content of stream sediment, much information was collected for other minerals besides uranium and the Survey decided to expand its exploration through the Geochemical Survey Programme. The principle of the programme is simple. By taking one sample of stream sediment from each square kilometre of the national grid, it is possible to create maps which illustrate the natural chemistry of Britain.

In practice, the success of the programme relies on a thorough sampling procedure and the British Geological Survey has reason to be pleased with progress to date. Scotland was completely surveyed by 1984 and in 1988 the geology teams had reached North Wales. Each summer of work adds a further 4000 items to the information bank of 60,000 samplings and each is analysed by

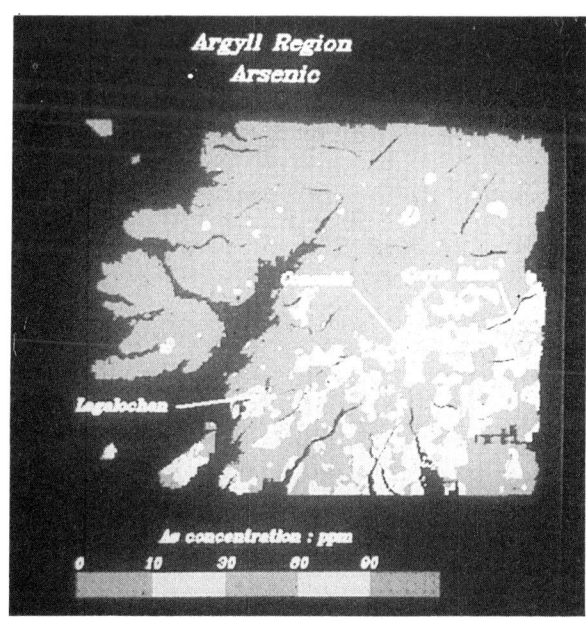

5. Image of the distribution of Arsenic over the Argyll region. The geochemical values delineate three principal target areas of Lagolochan, Cononish and Corrie Bhuie identified by Mining Companies during their exploration for gold. *(Extended caption on page v.)*

optical emission spectrometry for up to thirty different elements. The value of the work was enhanced when the Department of Trade and Industry funded a computer-based project called 'geochemical interactive systems analysis' (GISA) which involves the National Remote Sensing Centre at Farnborough.

By successfully combining GISA, the data from the survey programme and the Landsat images taken from space, it is possible to conjure maps of Britain on which are superimposed the locations of many requested elements. With such a system, there is no difficulty in displaying the gold locations but already "computer prospecting" indicates that a better plan is to study the proximity of those elements associated with gold discovery — namely, bismuth, arsenic and antimony.

In July 1989, The British Geological Survey announced discoveries of gold particles in rivers in Devon and Cornwall whilst working on a scheme funded by the Department of Trade and Industry. Dr Ramues Gallois of the BGS confirmed that an examination of satellite photographs and a study of chemical analyses had prompted the initial discovery which was then corroborated by geological students panning the Erme and Avon rivers in Devon, and at Wadebridge in Cornwall.

Dr Gallois' enthusiasm endorsed the Survey's methodology:

> We are pleased that we have been proved right in finding a group of rocks where we predicted there was a high chance of finding gold where no one had suspected its presence before. It looks like one of the better prospects in the United Kingdom, possibly the best in England.
>
> We are still only in the very early stages but our aim is to attract industry willing to invest tens of millions of pounds.

It was left to one of the local farmers to express a note of caution:

> I am not particularly pleased about it; we have enough trespassers. Mr David Balkwill, Aveton Gifford, Devon.

Reference
1. Prospectors pan back to the source Martin Belderson *New Scientist* 25th February 1989

PART I
ENGLAND AND WALES

ENGLAND'S GOLD

Because gold in Great Britain is most likely to be found on the line of longitude 4° West, England's opportunity to participate in the gold business is restricted to locations in Devon and Cornwall, and an ancient mine in Cumbria called 'Goldscope'.

The discovery of gold in South Devon was relatively recent. Whilst leading his students on a field trip concentrating on the Devonian rocks, in 1922, Professor W T Gordon of King's College, London, found gold on a small headland called Hope's Nose, near Torquay. In demonstrating how calcite would split in a predictable manner, the professor was surprised to find sprigs of gold binding the calcite together — "I had the good fortune to discover an interesting occurrence of gold in the fault-rock of a small fault cutting the limestone near Hope's Nose."

6. A rare specimen from Hope's Nose, Devon, which shows the delicate, fern-like structure of gold in its crystalline form.

Gold from Hope's Nose is very special because it occurs in the crystallised form. Etching away the host rock with a weak solution of hydrochloric acid reveals a delicate fern-like structure. This is a job for experts, and the ensuing specimens which are much prized by collectors are generally only available from dealers. In recent years, the Hope's Nose location has been nominated as a Site of Special Scientific Interest (SSSI) and unauthorised collecting is now against the law.

In North Devon, the copper mines at North Moulton were found to be carrying gold in the gossan ores in 1852. A test confirmed that there was more than one ounce of gold in every ton of ore and as it was of good quality, it sold for the then-high price of £4 4s. (£4.20) Within one year, over 130 ounces had been recovered and the owners of the Poltimore Mine were well pleased with their brief excursion into gold mining.

The traditional metal of Cornwall is tin but this does not preclude small quantities of gold being present in the same localities. A survey of Cornwall dated 1602 describes how the "tynners" would occasionally find small pieces of gold among their ore, which would be transferred to a quill for safe-keeping until there was an opportunity to sell it to a goldsmith.

Native gold has been found in most of the Cornish streams which flow to the south coast. Of these, the Carnon Stream at the head of Restronget Creek in the Falmouth Estuary has provided the greatest yields and it is not uncommon to find small nuggets there. In 1846 a scheme was initiated to mine for gold at Wheal Samson near St Treath but the results did not repay the effort and work soon ceased. A few years later (1852) the Geological Society of Cornwall reported gold-bearing quartz veins at Davidstowe in North Cornwall which, if nothing else, helped to prove that the gold of Cornwall is of local origin — rather than brought in by glaciation.

The gold mines of Goldscope in the Newlands Valley of Cumbria

7. Interior of the Goldscope Mine in the Newlands Valley, Cumbria, which can be explored in relative safety.

have an important place in history. By Royal prerogative, in England and Wales, all gold belongs to the Crown. In 1563 the Goldscope Mine was producing a lot of copper and a little gold. Queen Elizabeth I claimed it was a gold mine but the owner, the Earl of Northumberland disputed her claim. The ensuing legal wrangle lasted over fifty years and the concluding verdict established the law of ownership for all time. The final decision of the courts was that the gold would belong to the Queen, but not the other metals such as copper and lead.

Goldscope is said to derive its name from the German miners who came in 1566 to work the copper — a deposit which was so rich that they called it 'Gotzgab', meaning God's gift. By 1709 the name had evolved from Gowd-Scalp to Gold Scalp; the earlier presence of gold fully justifies the modern names of Gold Scoop and Goldscope.

Today the Newlands Valley belongs to the National Trust and the mine is easily accessible by footpath. It is possible to enter the mine and to explore in relative safety. Originally it was possible to walk through the mine so as to exit on the other side of the mountain after a journey of about 800 metres. Today there is little chance of finding any gold, but there are many mineral specimens to be collected in the interior.

GOLD OF SOUTH WALES

Following the conquest of Gaul (that is, modern-day France), Britain was invaded by the Roman Legions who landed on the south coast in AD 43 and spread their armies of occupation throughout the country. When the warring tribes of what is now Scotland presented a challenge to the Roman lines of communication, the Romans built Hadrian's Wall as a defensive measure and to mark the limits of their empire.

As part of the process of occupation, however, one legion made its way to a remote part of Wales known today as Pumsaint. The village is situated in hill country and the nearby town of Lampeter has been described, unfairly, as "lying between nowhere and nowhere". The Roman armies included soldiers with specialised trades and among the engineers were experts whose job was to locate and recover mineral resources such as copper, lead, silver and iron, which could be shipped back to Rome. It has to be assumed that their discovery of gold was a chance finding but there is no doubt that the Romans made the most of their opportunity. Their system of working, two thousand years ago, was different from modern times.

For the labouring tasks, the Romans recruited one thousand slaves; not having sophisticated tools, the engineers had a preference for working open cast. At Pumsaint, the mining operation involved digging out 250 feet of solid rock to create a man-made valley which is now known as 'the Ogofau Pit'. Because the mines all lie within an estate called Dolaucothi, and which has now been gifted to the National Trust, the term Dolaucothi gold mines is also used to describe the mines at Pumsaint. Although leading to confusion and ambiguity, it is acceptable to use the term 'Ogofau', which means 'a large cave', when describing the Roman workings and 'Dolaucothi' when considering the more recent mining activities.

Millions of gallons of water were needed to support their ambitious undertaking and, all around the area, the Roman engineers constructed a system of aqueducts to bring supplies of water overland. In one case the distance was seven miles. Quite apart from the processing operations, the Romans also used water for 'hushing' — a technique in which a controlled cascade of water rips off the top layer of soil and gravel to expose a new skin for prospecting. Another

8. The Roman adit at the Ogofau Gold Mine, Dolaucothi, South Wales, has a distinctive square section.

technique at which the Romans excelled was 'setting'. In setting, a fire is lit on the surface of a rock so as to raise it to a high temperature, whereupon water is thrown on the fire and the abrupt quenching action fractures and breaks up the rock.

It is a matter of conjecture how much gold was removed during the Roman activities at the mines. Some estimates establish a figure of one million ounces. On the other hand, by assuming a cautious return of about three grammes of gold per tonne of rock, the volume

9. Activity at the Dolaucothi mine, c 1933, at the time of the last initiative in producing gold in South Wales. *Reproduced by permission of the National Museum of Wales.*

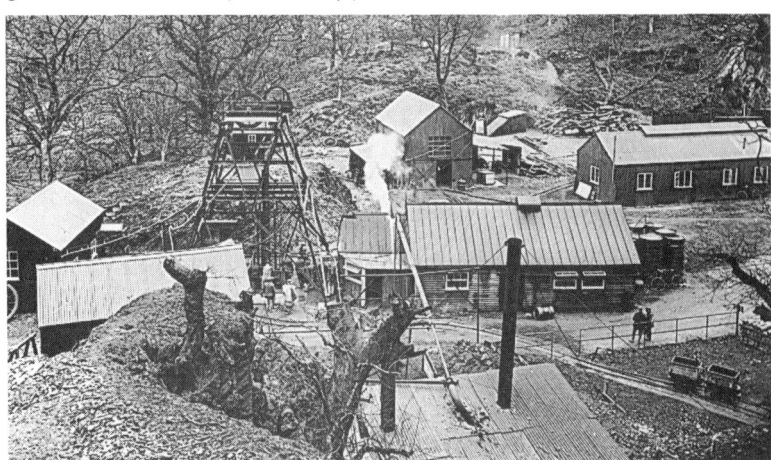

10. The entrance to Mitchell's Adit which was developed by James Mitchell in 1905 in his successful drive to make the Dolaucothi gold mines profitable.

of mass removed from the Ogofau Pit suggests a yield of about nine hundred kilograms.

After the Roman occupation, the mines lay dormant until the industrial boom in the nineteenth century which sparked off demands for coal, and metals such as copper, iron and lead. Gold was being discovered in other parts of the world and speculators were able to convince business investors that it was reasonable to expect a return from prospecting for gold in Wales. As a direct result, the short-lived Eurion Eurglawdd Mining Company was established at Dolaucothi in 1871, to be followed by South Wales Gold Mines Ltd in 1888. Although gold was found, it was a lack of profits on the venture which led to work ceasing. Fortune improved in 1905 when the owners of the estate employed James Mitchell who had experience of the South African gold mines.

Mitchell was able to prove that investment would pay and he was granted a lease from which the Ogofau Proprietary Gold Mining Company was born. Mitchell continued to make a small profit, and encouraged, he drove shafts and adits further into the hillside. 'Long Adit' measures 400 metres and connects with an internal vertical shaft derived from the Mitchell Adit, which is to be found at the top of the hillside. Unfortunately his capital was exhausted before he struck it rich and he had to sell his stake to Cothy Mines Ltd in 1909 and work for the new company as its mining manager. It was during this period that Mitchell explored even further and dug downwards to a depth of 100 feet where he encountered a quartz lode rendering just less than 1/2 ounce of gold per ton. In spite of his success, however, flooding and difficulties with pumping forced yet another closure on the operations.

From 1912 the mines remained closed until a fresh initiative was begun by Roman Deep Holdings Ltd in 1934. Using methods of the twentieth century, Mitchell's ambitious shaft was extended down to 260 feet from where a further 400 feet of probing confirmed ore reserves with gold in paying quantities. To take advantage of the situation, a new company, British Goldfields (No. 1) Ltd, was formed with one hundred men and the main shaft was extended yet again to 480 feet.

Much of the gold at this depth was concentrated in sulphides with a high arsenic content. This called for processing in specialised plants which were located in the United States of America and in Hamburg. The transportation costs to America ruled out the possibility of profit, but by treating the auriferous material in Hamburg, there was a small return. However, political pressure and the impending outbreak of war (in 1939) brought about the end of the golden age at Pumsaint. There is no reason, however, why it should be anything more than a temporary setback.

Nowadays the mineral rights to the gold belong to the Crown, but the gold mines are in the possession of the National Trust — and effectively belong to the nation. In an ingenious arrangement between the National Trust and the University of Wales, research and restoration is conducted by the Department of Mineral Exploitation at Cardiff, and the Archaeology Unit at Lampeter. In the summer months, the two departments combine to present a happy blend of interpretations and mine tours. Under the guidance and supervision of students studying mining and archaeology, Ogofau has emerged as an important tourist amenity and one which reveals the evidence of the Roman occupation very much as the Romans left it.

To the south of the gold mines and within easy walking distance, there is a sequence of Roman adits which have been tunnelled from solid rock and are, therefore, relatively safe to explore.

Equipped with protective clothing and torches, it is possible for visitors to view the delicate chisel marks and the dressed stone of the Roman miners and to wonder at the sections determined by their engineers. Lower Roman Adit is coffin-shaped. That is, it is wide at the top and narrow at the base so as to ease the carrying of loads on the back. Nearby, Upper Roman Adit is square sectioned and penetrates the hillside to a distance of about 250 yards before it opens out to a large chamber revealing a collapse and a small shaft through which it is possible to scramble out to the open air. It has been suggested that the square shape of this tunnel was to provided space for two teams of workers — one group entering and the other group leaving.

There may not be the gold to satisfy the urge of the recreational gold prospector but if he has to take a rest from panning, the gold workings at Ogofau in South Wales provide the perfect setting for an archaeological study of an earlier age in gold history.

THE DOLGELLAU GOLD BELT OF NORTH WALES

Following the chance discovery of gold near Dolgellau in the middle of the 19th century, North Wales has emerged as an important British location in view of the commercial mining which has been carried out over the past one hundred years. Alas, no great fortunes have been made, but in view of the regular interest created by royal wedding rings which are made from Welsh gold, a well-deserved reputation has been established around the world.

It was in 1843 when galena was being mined on the east bank of the River Mawddach at a small mine called Cwm-heisian that the company consultant, Arthur Dean, pointed out that gold was accumulating in the machinery being used for dressing the lead ore. It did not take long for the production to switch to gold recovery, and to establish a place for the Cwm-heisian mine in history.

There is not much to see at Cwm-heisian today even although in the heyday, mining was carried out at two places about one thousand yards apart. The enthusiast will find a number of adits, the pit of a large water wheel and a protected shaft — all partially obscured by the dense forest. Of more interest is a small processing plant which was built to serve both sites on a small and flat piece of land, and its remains are an important feature of any walk along the Gold Road which follows the River Mawddach from the A470 roadway to the Gwynfynedd Mine.

During the year when gold was discovered at Cwm-heisian, gold was also located at Dolfrwynog for the first time. The circumstances leading to the discovery were unusual and originated when a visiting geologist reminded villagers that it would be beneficial to spread the ashes from their peat-burning fires on the fields. On being assured that to do so would kill the plants, he asked for a sample of ash which, on analysis, was found to be very rich in copper. It was not long before some enterprising individuals reasoned that the mineral deposits might extend to gold and, as a consequence, many trials and tests were carried out in the neighbourhood. Gold was eventually located and in the interim, the mine became known as 'The Turf Mine' in view of its unconventional system of metal recovery. That is, by burning procedures rather than smelting.

11. The Turf Mine at Dolfrwynog, where there is little to see, but which is the site of one of the two simultaneous discoveries of gold in North Wales in 1843. *Photograph by Andrew Callender, December 1988.*

At the site today, one can see the remains of a small building which includes a suggestion of a furnace, some ponds set into a rivulet and a large stone which could have served as a claim marker in earlier days. Nearby are shafts, adits and spoil heaps; all require the benefit of doubt in their interpretation.

It was some years later before gold was found in substantial amounts in North Wales; 1860 is generally accepted as the date of the Welsh Gold Rush following an important discovery of the yellow metal at the Clogau Mine, near Bontddu on the Barmouth Road. Production was maintained at this mine under a variety of companies until 1986 when mining ceased in order that the mine could be developed as a tourist amenity.

An important aspect of the 1860 find is that it led people to accept the notion that gold could be found in Wales and not be dismissed as fantasy. This marked the beginning of a serious phase and during the next decade, mines and mining companies sprang into being at many places and it has become acceptable to describe the area as The Dolgellau Gold Belt. This gold field encompasses an area of about twenty miles, which stretches from Bala in the east to Bontddu in the west, with the Prince Edward Mine at Trawsfynedd forming the northerly edge. Contained within the belt are at least fourteen well-known mines, and old names, Welsh names, romantic names and Royal names have all been used to identify them. Over the years, some have changed name as their fortunes altered. When Bedd-y-coedwr provided gold for the wedding ring of the Duchess of Kent in 1934, the mine earned the right to a new title — the Princess Marina Mine. At one time Cefn Coch was known as the New Californian Gold Mine, and when gold from Bwlch-y-llu Mine was used for the regalia at the investiture of the Prince of Wales in 1911, the mine received permission to change its name to The Prince Edward.

However, it should be remembered that most of the mines were small operations run by a few men and subject to regular changes of ownership. Two exceptions were the Clogau Mine and the Gwynfynedd Mine which each had their origins in the 19th century but managed to carry on commercial mining into the 1980s.

The Clogau Mine is about two miles from the Vigra Mine which is better known for its copper but when the initial gold strike occurred at Clogau in 1854, Vigra promptly re-equipped its processing mill so as to handle gold ore. By good fortune the mill was centrally placed and ore was transported to it by railway, by incline ropeway and, from the Tyn-y-Cornel level, by an aerial system of buckets. The Tyn-y-Cornel level was begun in 1880 as a cross-cut by the Clogau

12. The No. 1 vein (St David's vein) of the Clogau Gold Mine which was actively worked during the period 1859 to 1862. Protected nowadays by a fence, it would be foolhardy to attempt an unsupervised exploration.

Mining Company in order to reach the main lode which is known as the St David's No 1 vein.

For ten years the return on the investment was minimal and it fell to a new company — the Clogau Gold Mining, but often known as 'The Welsh Company' to make the operation profitable. Some years later, (1898), the Welsh Company was replaced by The St David's Gold and Copper Mines, and when the latter joined forces with The St David's Gold Mines company in 1903 (in an endeavour to marry a partnership between the operations at Gwynfynedd and Clogau), the new initiative led to a sharp increase in fortune.

This was long overdue and production increased by a factor of three. Out of the prosperity came a decision to drive a new adit at Llechfraith onto the lode in 1904. The official record, 'Mineral Statistics', confirms that 1904, 1905 and 1906 were important years for Clogau Mine, but sadly, the output was not maintained in subsequent years and by 1911, mining had ceased. It is worth noting that the Llechfraith Adit is an important feature of the plans to develop Clogau as a tourist centre over the next few years.

The Gwynfynedd Mine, by the banks of the River Mawdach, shares an equal place in gold mining history with Clogau. Over the years its name has changed to include The Gwyn Mine, Mount Morgan, and British Gold Fields. It was the discovery of a very rich shoot of gold which first launched the mine in 1887; a stroke of luck which sustained work up until 1905. From then on, mining was continued with varying success until 1917, but sporadic interest in subsequent years meant that production ceased on the outbreak of war in 1939.

A bold initiative took place in 1985 when Mark Weinberg and two partners decided to open the mine once again. The mine had lain derelict for many years and in spite of economic and other hardships, the venture managed to keep going until 1989.

The destiny of the Gwynfynedd Mine relies on the jewellery trade and the reasoning that customers will pay a premium price for 'Welsh Gold'. To this end, the special mark of the Welsh Maiden has been revived, and many wedding rings carry the image as proof of authenticity. Rumours claim that if extra men were engaged (on a third shift) to work around the clock, the mine would be profitable, but environmentalists point out that the extra activity would increase the 'spoil' (mine debris) which would seriously harm the countryside and, in particular, the ecology of the river. It would be no surprise,

MAIDEN WELSH GOLD.

Gold has been produced in Welsh Mines since 1850. However its relative rarity has made pieces containing Welsh Gold collectors items and added value to jewellery. Traditionally Gold produced in Wales has supplied Rings and other Royal items. Some examples are the Wedding Rings used by Prince Charles, Princess Diana and the Queen Mother.

13. *(above, left)* Cover and 14 *(left)*, centre of Certificate of Authenticity which accompanies 'Welsh Maiden' gold products. 15 *(above)*. A billet of gold from the Gwynfynedd Gold Mine carrying the authenticating trade mark of the 'Welsh Maiden'.

therefore, if the Gwynfynedd Mine ceased production and reverted once again to an abandoned mine.

It is small consolation that in North Wales, most of the abandoned mines are reasonably accessible and are worth visiting. For example, Cefn Coch and Berthllwyd Mines are both on National Trust property on the east slope of Garn Mountain near Ganllwyd village. A strategic set of signposts indicates the pathway for a circular tour of the site. At the base of the mountain there remains the foundations of a processing plant which includes a fine example of the Arrastra — a device for crushing rock. From here an old railway track leads to the Berthllwyd Mine entrance. Nearby is the inclined trackway of an early transport system and a stiff climb along it leads directly to the portal of the main adit of the Cefn Coch Mine.

16. A group of 19th century gold miners at the end of a day shift which includes Hugh Pugh (*lower right*) whose diaries provide details of contemporary working conditions. *Reproduced by permission of Gwynedd Archives and Museums Service.*

17. The Tyn-y-Cornell level at Clogau Gold Mine which was cut in 1880 in order to simplify access to the main gold veins.

18. The main adit of the Cefn Coch Gold Mine on Garn Mountain, near Ganllwyd.

19. Interior view of the Gwynfynedd Gold Mine at the heyday of its commercial activity in 1986.

20. Forlorn symbol of the Gwynfynedd Gold Mine which ceased operations in 1989 after a burst of activity which had lasted five years.

Close by are the spoil heaps and, for the enthusiast, there are more adits, levels and open cast working further up the mountainside. But it is not necessary to venture underground to appreciate the life and times of the Welsh gold miner. The main adit is linked by a small-gauge railway track — which survives in places — to the substantial remains of another processing plant, thought to date from 1875. Nearby are the explosives store, settling ponds and the barracks which housed the miners. But Cefn Coch provides a final bonus . . . the rewarding views of the Mawddach valley and the opportunity for good photographs of a chapter of nineteenth century history.

By necessity, this description has been brief but for further reading, the references overleaf are recommended.

Notes
1 The following list of gold mines has been compiled from the book by Morrison and is included here for completeness.

Cwymprysor	Prince Edward	Bedd-y-coedwr
Cwm-heisian	Gwynfynedd	Tyddyn Gwladys
Dolfrwynog	Cefndeuddwr	Ffridd Goch
Glasdir	Cefn Coch and Berthllwyd	
Cae Mawr	Caegwernog	Wnin
Cesailgwm	Moel Ispri	Prince of Wales
Cambrian	Carmabseifion or East Clogau	
Garthgell	Clogau	Vigra
Graig-wen or Nant-goch		Panorama

2 It is sensible to take precautions before entering any abandoned mine, and it is best to investigate under the supervision of a person with experience. As a general rule, there ought to be at least three persons in the party and all should be equipped with hard hats and

torches, as a minimum. In addition, an old anorak and rubber footwear are useful provisions.

References
1. *The Gold Mines of Merioneth* by G W Hall and published by the author from 17(a) Bridge Street, Kington, Herefordshire HR5 3DL.
2. *Goldmining in Western Merioneth* by T A Morrison is now out of print but was originally published by Merioneth Historical and Record Society in 1975.
3. *Treasures of the Mawdach* by Hugh J Owen includes a chapter on gold mining which is a good contemporary account of the events by an eye witness.

PART II
SCOTLAND

SCOTLAND'S GOLD

CRAWFORDMOOR

Crawfordmoor is an ancient title describing goldfields in Southern Scotland which stretch from Mennock Water in the west to Meggat Water in the east and which embrace the lead mining villages of Leadhills and Wanlockhead in Lanarkshire and Dumfriesshire. Credit for the discovery of gold mines in Scotland is always accorded to King James IV who raised the finance* in 1511 for Sir James Pettigrew to prospect on his instructions. Just as the King's mines were becoming profitable, he was killed at the Battle of Flodden and, for a time, the mines were controlled by a Queen Regent who directed the output to the Scottish Mint in Edinburgh for conversion to ducat and unicorn coins.

In 1526 leases were granted to Dutch and German adventurers, and following a commission of enquiry in 1535, miners from Lorraine

21. A map by John Blaeu dated 1635 features Crawfordmoor and locates the area just south of the river Clyde. *Reproduced by permission of the Trustees of the National Library of Scotland.*

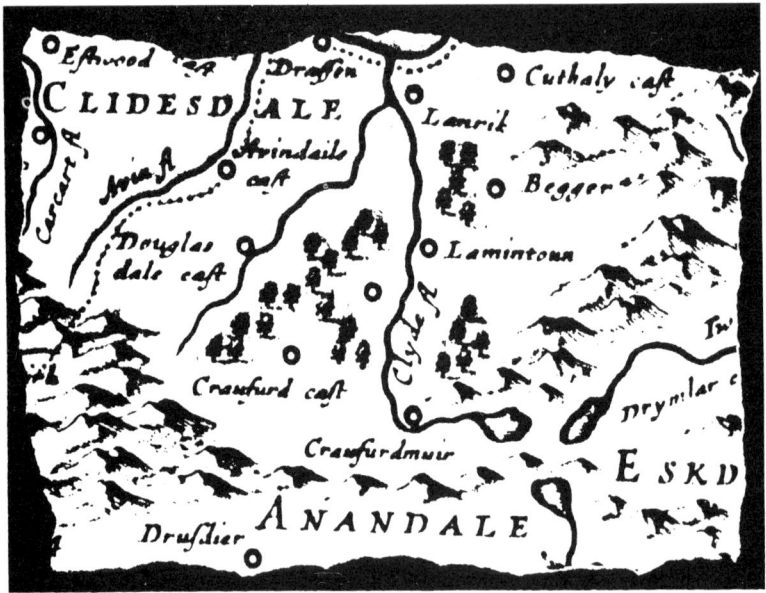

were permitted to work the gold deposits. By all accounts, much gold was extracted and removed from Scotland by these exploits - in spite of a law forbidding the export of native gold. When King James V annexed the throne, he ensured some gold would remain in Scotland for all time by commissioning a crown for himself and his queen. The origin of the king's crown is obscure and it is probable that an earlier crown was refashioned for James. What is certain is that the arches were added at this stage and it was at his insistence that "the king's own gold" should be used.

The mining ventures had a renewed burst of activity in 1567 when licences were assigned to groups of prospectors on condition that a

22

23

22. It is permitted to sketch the Honours of Scotland provided no session extends beyond two hours. Drawing of the Scottish Crown made by the author in the Crown Room, Edinburgh Castle, April 1984.

23. Crown copyright controls illustrations of the Honours of Scotland, but the Scottish crown was used on an insurance plaque in the 19th century which anticipated the protection by law.

24. A Scottish law of 1524 prohibited the export of gold in order that it could be converted to ducats at Cunyie House in Edinburgh. The coins featuring King James V became known as 'bonnet pieces' in view of the 'capberet' which the king favoured in his portrait.

24

21

royalty of around six per cent would be paid on results. Cornelius de Vos formed a syndicate in Edinburgh and raised £416. He recruited over one hundred persons to his work force:

> ...he employed both lads and lasses, idle men and women, which afore went a-begging, and he profited by their work, and they lived well and contented.

The rate of pay was fourpence a day and a frequently quoted story tells how 138 ounces were produced in 38 days. The contemporary value is quoted as £450.

Crawfordmoor's most famous prospector was Sir Bevis Bulmer who arrived in 1578 with letters of authority from Queen Elizabeth and from King James VI of Scotland "to make and adventure, and seeke for the gold and silver mines". He had boundless energy and prospected all the rivers in the district — Short Cleuch, Glengonner, Lang Cleuch and Wanlock Water.

> On Wanlock Water he caused diligently search for natural gold, of which he got a pretty good quantity, and made watercourses to wash it . . . . By help of the watercourse he got much straggling gold, on the skirts of the hills; and in the valleys which kept him in great pomp, keeping open house for all comers . . . .

Some of Bulmer's nuggets weighed five and six ounces and his wash-heaps can still be seen at the confluence of the Lang Cleuch and Elvan Water. On the summit of Bulmer Moss, which dominates the valley, it is possible to locate features which match the ancient description of his "surface diggings that cover the flanks of hills".

A contemporary of Bulmer was George Bowes who was financed by Queen Elizabeth in order to find the source of the gold. At the end of 1603, he claimed he had "discovered a small vaine of gold which had much gold in it". Bowes convinced his workmen of the need to keep the secret and "never to disclose same unto the King of Scotland, nor his Counsell" and when he reported to his patron, the Queen of England, she made arrangements for him to return the following spring to continue his work. Carelessly, Bowes fell down the shaft of a copper mine in Cumbria and was killed. His "secret" died with him.

For the next one hundred years there was little activity on the gold scene. Charles I had coronation medals made from Crawfordmoor gold and plans were drawn up in 1740 to work the mines. This and

25. Reproduction of the first folio of the Cottonian manuscript (now fire damaged) which is a 17th century account of the gold mining activities on Crawfordmoor. The report begins:
   First. I have been informed in King James (IV's) time, Scottishmen did begin to wash gold, and in some summers, there was three hundred persons which did maintain gold; but for this forty last years there hath been little washing but in the foresaid eighty years many gills, waters and valleys have yielded gold therein, of greater value than an hundred thousand pounds; yet by the people working for gold no veins of gold have been known to be found. *Reproduced by permission of The British Library,*

another scheme in the reign of George III also failed to materialise. By now the lead mines were in full production and it is fair to say that the political and economic climate was not favourable to gold prospecting.

26. Design of the keeper ring made from local gold which was presented to Princess Mary of Teck in July 1893. From the archives of the Miners' Library, Leadhills.

27. A special photograph of Princess Mary of Teck displaying the gold ring which was presented to her by the miners of Leadhills on the occasion of her marriage in 1893.

A second golden age occurred in the late 19th century. Although the lead mines still held sway, many of the lead miners were first class gold washers. In their spare time, they would supplement their earnings by undertaking commissions from Victorian and Edwardian mineral collectors. The men used a "trough" which was a type of sluice box placed in the river to wash the gravel by the skilful control of water passing through a gate.

The semi-professional approach to gold prospecting continued

28. John Blackwood was an active prospector from Leadhills village who discovered "the biggest nugget found in Scotland for two hundred years" whilst working on the Windgate Burn on 6th June 1940. A religious man, Blackwood exclaimed, "Lord, don't let this gold get a hold on me". The illustration shows him using a trough near the 'Gold Scaurs' by Elvan Water.

into the middle of the twentieth century when there was a sensational find in 1940. A report in *Scottish Field* by Harry Hutchinson gives the flavour of the occasion:

> On June 6th, when the attention of everyone was focussed on Dunkirk, Mr John Blackwood, prospector and retired lead-miner, was searching for gold in the shingle beside the Wyngate Burn, which runs from the Green Lowther, when his pick dislodged a piece of quartz about as large as a pullet's egg. As it fell heavily towards the bottom of his 'place' he glimpsed a yellow colour. He stooped quickly to retrieve the quartz and, greatly excited, he noted that on top it bore a solid tongue of gold. John Blackwood stuffed the nugget in his waistcoat pocket and hurried home to examine his find. The mass of the lump of gold-bearing quartz was some 500.2 grains, of which more than half was pure gold, making it the biggest nugget found in Scotland for more than two hundred years.

Today's gold prospectors on Crawfordmoor are recreational and apply themselves to the task in order to have some good, outdoor

29. "Prospectors" pose for a snapshot on Wanlock Water, c 1928. It is thought that two gold miners have been interrupted by a passerby who has posed for the picture. Reproduced by permission of the Administrator of Wanlockhead Museum Trust.

exercise combined with an urge to maintain a tradition based on gold and the Scottish goldfields. It is not an easy pursuit and no great fortunes are to be made. The famous rivers of history — Windgate,

30. A modern day prospector working on Mennock Water, May 1985.

31. Carved stone slab over a doorway in George Heriot's School, Edinburgh, which shows a 17th century goldsmith's workshop. *Reproduced by permission of the Governor's of the George Heriot's Trust.*

Lang Cleuch, Short Cleuch, Elvan Water, Wanlock, Glengonner and Mennock Water cover such a wide area that it is not easy to switch locations on a whim. The weather can make a difference to one's fortune. After a drought, reduced water levels give access to parts of the river which are inaccessible in normal circumstances. On the other hand, heavy rainfalls help to liberate more gold from the ground and flush it into the rivers where it will travel downstream until it is trapped in a crevice or rests behind a rock. At times the water can be very cold and it is necessary to dress appropriately in view of the strenuous nature of panning.

Gold can still be found in small quantities and it was with remarkable foresight that Stephen Atkinson wrote in 1619:

The same gold mines will become to be an
everlasting happiness to all successive ages.

Amen. Amen.

Reference
1. *The Discoverie and Historie of the Gold Mynes in Scotland* Stephen Atkinson (1619) James Ballantyne & Co. Edinburgh, 1875.

THE GOLDEN ZONE

For years enthusiasts have shared knowledge of the gold-bearing rivers of Central Scotland and exchanged maps of the Cononish river, near Tyndrum. Well known in the 18th century for its lead mines, in 1985 the Ennex Corporation declared its interest by conducting explorations for gold in Beinn Chuirn mountain. In 1986, the Dublin based company completed a reconnaissance for gold deposits and confirmed 'showings' at seventeen locations. Further exploration continues with the expressed objective of bringing this 'geological reserve into production''.

In the meantime, the Scottish newspapers carry regular accounts of the progress being made and publish interviews with Mr John Burton, the farmer on whose land the activity is centred. Burton stands to benefit from renting his ground, and the Crown will levy a commission of 2% of the takings. The actual mining is expected to start "some time in 1990", says Ennex geologist, Richard Parker, "There is certainly enough ore to make mining worthwhile but I am not prepared to say just how much money the mine will generate, because these figures can be misleading."

Regular reports, however, estimate that the deposits could be worth £50 million. This view is confirmed by Dr Michael Gallacher of the British Geological Survey office in Edinburgh, "There is no doubt mining in Scotland and Ireland has taken off. I expect there will be

32. Disclosure of the 'Golden Zone' in *The Sunday Post*, 11th December 1988.

33. Press cuttings which proclaim the current gold bonanza at Tyndrum.

other finds but it will take detailed scientific work and a lot of hard grind, but it is excellent news for the Highlands. We have not had gold finds in Scotland like this before. In world terms the finds are not major, but in the European league table they are important."

In December 1988, another surprise announcement disclosed that the Colby Resources Corporation of Vancouver, Canada, had been conducting test bores at Aberfeldy — which is about thirty miles east of Tyndrum. The corporation is able to forecast gold mining to a

34. Panning in the waters of the Cononish River with the snow-capped Ben Lui in the background.

29

value of about five tonnes a year which would put Scotland on a par with Europe's biggest producer, France. It was earlier reports which had hinted that a geologist working for British Petroleum had traced gold deposits at Kilmelford and this gave rise to the description of the fifty miles "golden zone" of Scotland.

THE KILDONAN GOLD RUSH

Following the discovery of gold in California in 1849, the yellow metal was found at many other places around the world during the next forty-five years. Scotland ensured its place in the history books late in 1868, when a brief announcement in a local newspaper stated that gold had been discovered at Kildonan in the county of Sutherland. The credit for the discovery goes to Robert Nelson Gilchrist, a native of Kildonan, who had spent 17 years in the gold fields of Australia.

On his return home, he was given permission by the Duke of Sutherland to pan the gravels of the Helmsdale River and he chose to prospect all the burns and tributaries in a very methodical manner. He found gold in many places but the greatest concentrations were in the Suisgill and Kildonan Burns. The accounts of his findings spread like wildfire throughout the north of Scotland. *The Illustrated London News* circulated the story further afield and, within six months, over 600 hopeful adventurers had made their way to the

35. Portrait of Robert Nelson Gilchrist who discovered gold at Kildonan in 1868, which has been assembled from contemporary descriptions, and information provided by his family.

normally deserted Highland Glen. Two local papers, *The Northern Ensign* and the *John o'Groat Journal* carried regular reports "from the diggings", and *The Inverness Courier* and *The Scotsman* also maintained a regular stream of news from their "special correspondents".

In those days, the railway line terminated at Golspie and the last thirty miles had to be tackled on foot. Many of the prospectors were novices but a hard core of miners from Australia and America helped to provide some much-needed expertise in gold recovery. In April 1869, however, the Duke of Sutherland introduced a system of licences which cost one pound per month for each claim measuring forty square feet. In addition to this, the prospectors were expected to pay a royalty of 10% on all gold found; not surprisingly, much gold was never declared but was used in barter for food, tools and accommodation.

By this time, two small 'towns' had come into being. Baile an Or was a settlement of huts which was established by the banks of the

36. Announcement of the discovery of gold in Sutherland which appeared in *The Northern Ensign* on 10th December 1868.

NORTHERN ENSIGN, THURSDAY, DECEMBER 10, 1868.

DISCOVERY OF GOLD IN SUTHERLAND.

No small sensation has been created throughout the eastern district of Sutherland within the last few days by a report that gold had been found in Kildonan Strath; and though it has not yet so affected the public mind as to lead to a 'rush to the diggings,' it has raised sufficient interest to justify such an amount of 'prospecting' as will put an end very soon to all doubt on the subject.

Meantime it has been established beyond doubt that gold is to be found there, and the only question for solution is as to the quantity. We believe that in addition to gold, there has been detected the presence of scarcely less precious minerals in the same locality, and that a very decided impression prevails that considerable quantities of both descriptions may be obtained. No time will be lost in testing the matter. Meantime the fact of gold being in Kildonan is established by the successful search of several individuals, and the report of mineralogists as to the quality of the ore.

MYSTERIOUS CASE OF DROWNING AT WICK.

A PAINFUL feeling has been created here this week by the drowning, in Pulteneytown harbour, of Malcolm Green, mate on board the 'Aucheen,' of this port, now lying

№ 1176 № 1176 **SUTHERLAND ESTATE.**

PARISH OF KILDONAN.

1869.

Gold Miner's License.

License to _____

BY HIS GRACE GEORGE GRANVILLE WILLIAM, DUKE & EARL OF SUTHERLAND, K.G., to _____ (hereinafter called the Miner).

In Consideration of the sum of _____ pound _____ shillings and _____ pence, paid to the said Duke and Earl by the Miner, the receipt of which is hereby acknowledged

Residence _____

This License, subject to the conditions endorsed hereon, and to the payment of the Crown Royalty therein mentioned, is granted by the said Duke and Earl, signed by his Factor on his behalf duly authorised, to the Miner, to dig and search for Gold, for his own use and behoof, in the alluvial deposit along the sides of the Kildonan, Suisgill, and Torish Burns, and tributaries thereof, in the Parish of Kildonan, from and after this date, until the expiration of the present Month of _____ when this License shall terminate.

County of _____

_____ days.

£ : :

Dated this _____ day of _____ 1869.

Entd. _____ _____ *Factor.*

N.B.—This License to be produced on Payment of Crown Royalty.

37. Facsimile of the licence which was introduced by the Duke of Sutherland by proclamation on 29th March 1869. Obtainable from a licence office run by an inspector, it cost one pound (in an age when £8 was the average annual salary) and permitted the gold miner to work for one month provided he paid a 10% royalty and, if appropriate, a tent fee.

Kildonan Burn and Carn na Buth (meaning Hill of the Tents) served the workers on the Suisgill Burn. Very soon a 'saloon' was added to the Town of Gold (Baile an Or) and provided meals and accommodation for the mining fraternity. *The Northern Ensign* carried a description of the stores:

"One end is devoted to culinary, kitchen and lodging purposes, the beds being arranged in layers like those of a steam boat. Potatoes, salt, butter, eggs, bread, cheese, onions, cabbages, spade handles, tobacco, pipes, snuff, writing paper, stockings, shirts, tea, sugar, barley, oilskins, candles, beef and castor-oil, are all arranged in the most admired confusion, and are to be had at prices which might disturb the equanimity of local provision merchants and dealers in castor- oil to quote."

During the summer months of June and July the issue of licences continued at levels of about two hundred per month. Many tourists and journalists arrived by the stage coach which ran from nearby Helmsdale and the Duke of Sutherland paid a visit in order to present a valuable gold watch to Robert Gilchrist as the "discoverer of Sutherland gold". A drought in June allowed the diggers access to the gravels on the river bed and this may account for the fall in the gold price. *The Scotsman* reported "it can now be got at £3 10s

38. The 'town' of Baile an Or (Gaelic : Gold Town) which *The Inverness Courier* described on 29th April 1869:
> Most of the houses are wooden huts, neatly and cleverly put together. Here and there among them appears a canvas tent of unmistakeable Australian design, having its double roof, alike to keep out the rain and excessive heat of the sun, while the beds in which the miners sleep are raised several inches from the ground. Butchers, bakers and grocers' shops are established and seem to be doing a lucrative trade. A gold office is also to be seen, with a notice outside, intimating that 'Gold is bought here'.

39. On the 2nd June 1869 the Duke of Sutherland presented "a valuable gold watch" to Mr Robert Gilchrist to mark his discovery of Sutherland gold. In a report in *The Northern Ensign*, the Duke is described in some detail:
> His Grace is not august or regal; he is not a simpering puppy or a swaggering ninny; he is not great in gold chains or valiant in neckties; he is a good-looking, handsome young Highlander, having no more affectation than broad cloth but clad in a good suit of grey tweed, red shirt, Scotch bonnet, coarse shoes and sound, sober mind. A noble looking fellow, with first rate legs . . . .

THE SUTHERLANDSHIRE GOLD DIGGINGS

KILDONAN BURN

BAILE 'N OIR, KILDONAN

40. News of the Kildonan Gold Rush spread to England on publication of an account of the Sutherlandshire Gold Diggings in *The Illustrated London News* for 29th May 1869. Reproduced by permission of the The Illustrated London News Picture Library.

41. Advertisement in The Inverness Courier for July 22nd, 1869 offering all the supplies for the gold miner's needs as well as opportunity for visitors to purchase gold. Note that this notice acknowledges Baile an Or as a placename.

(£3.50) per ounce and when buyers are not plentiful, sales are being effected at £3 8s (£3.40)." [Note that at the start of the gold rush the going rate was as high at £4 10s (£4.50).] Nevertheless sales of 1.5 to 2.5 ounces per week were considered "fair wages" and many men expressed satisfaction in being able to wash eight or ten shillings worth of gold each day. That is, sums of value 40 to 50 pence.

The start of the herring season in August depleted the numbers and by September, the number of miners was down to fifty who, with some justification, were anxious for the Duke of Sutherland to

42. Announcement signalling the end of the Gold Rush.

43. A downstream view of the steep-sided canyon of the Kildonan Burn.

allocate more ground for gold prospecting. The Duke's reluctance, however, was on account of complaints from his tenant farmers and the fishermen. As the weather deteriorated, the outlook appeared bleak and it was no surprise when the diggers were summoned together to hear an order read by the Inspector. "After Saturday first, no new licences will be given out and on the expiry of those current, the diggers must remove their tents and leave the locality." Applying the principles of market forces, the diggers now demanded and received £4 an ounce for their gold. The Scottish Gold Rush ended at midnight on the 30th December 1869.

AFTER THE GOLD RUSH

There was a fresh initiative from Sutherland County Council in 1895 when the gold diggings at Kildonan were opened for a serious exploration. The equivalent of 476 man-days produced only thirteen ounces of gold and the experiment was deemed to have failed. Mr Seaton, the local stationmaster, had been put in charge and for a number of years there were accusations that the tests had not been conducted fairly. The gold diggers had been restricted to one piece of ground and not only had the tools been inadequate for the task in hand, it was said that not all their earnings had been passed over to the stationmaster.

For political reasons, the son of the Duke of Sutherland, the Marquis of Stafford, permitted more tests in 1911. Mr Heath, "an expert with 13.5 years experience in the Yukon", was appointed to take charge of the operation. "Physically and intellectually strong, he is a man to inspire his workers with confidence, and will control them with firmness, courtesy and kindness." Local men were engaged and a substantial camp was prepared "in a quiet spot a short distance above the bridge over the Suisgill Burn". The work was planned to start on Monday 26th June 1911 but was delayed. "The manufacturers of the canvas required for the tents have not been able to deliver the material on account of pressure of Coronation work". "Breakfast is to be at six. Work begins at seven; there will be an interval for dinner at one, and we finish off at six. I am to give the men a square deal", Mr Heath is quoted in an interview published in the *Northern Times* for 15th June 1911.

44. A view of the upper reaches of the Kildonan Burn which shows an elaborate dam and three men operating a sluice box.

45. A rare photograph of the owner of the Suisgill Estate in 1914 supervising a sluicing operation. Mr Hirscht had arrived some years earlier from Germany with half a crown (12½p) but was interned on his estate throughout World War I.

By August, "after weeks of hard toil", it was reported that "Mr Heath and his men have now struck a small paying streak of ground in the Gold Burn. In the Suisgill Burn, trenches have been cut at several places on both sides of the stream, the gravel has been thoroughly sifted and the sand in the hills has been carefully examined."

Alas, when Mr Heath presented his report on 16th October his conclusion was to the point:

> After carefully considering all the facts that these operations have determined, I regret to be forced to conclude that this field cannot be worked by any method so as to realise on the most liberal estimate anything like half the actual working expenses.

With a budget of £2000, in 1969, Sutherland County Council commissioned the Institute of Geological Sciences to carry out a survey of the Strath of Kildonan. A further grant from the Highlands Development Board the following year provided for trenching and test bores in the alluvium. The final report was a familiar story:

> Results show that gold, though of widespread distribution, is not concentrated in economic proportions in the alluvium. The geological and bore-hole evidence suggests that the disseminated gold in the Strath was liberated in an early tropical erosion cycle and redistributed by ice action.

LETTER FROM INVERNESS

In 1987 the author traced a letter which had come from a young man from the South who had been asked to write down "what he had seen and knew in reference to the goldfields at Kildonan". The following edited extracts present a realistic eye-witness account of the gold mining community of 1869. The letter is dated 17th July 1869.

"I shall endeavour to dissipate the false halo of romance in which the Sutherland gold fields have been partially enveloped, owing to the glowing and exaggerated reports which have found their way from time to time into the columns of the Press.

"Without further preamble, I hold that the Sutherland gold diggings have as yet, for all practical purpose, proved a most complete failure. It is a fact that with the exception of a few poor fishermen who, owing to the introduction of the licence system, are now precluded from digging, few, if any, have made their own of the Scotch gold diggings. True some dozen shop keepers, a publican or two, and three or four storekeepers, all more or less interested in the circulation of inflated reports of them have benefited by the diggings.

"More than one instance has come under my own direct observation of men who left their families, travelled a long journey — provided themselves by dint of severe pinching with expensive tools — paid £1 a-month for permission to dig forty square feet of rocky moor — worked like galley slaves month after month — and in the end were obliged to sell their all in order to provide themselves with the very necessaries of life as they trudged miserably home on foot.

"Arrived in Helmsdale from Golspie after a walk of 18 miles, I lose no time in procuring cheap lodgings, for the day had been excessively hot, and I was tired and footsore. Having but ten miles before me 'ere I reached the land of gold, I am up with the lark on the morrow. My way lay along a dreary and uninteresting strath, through which flowed the river Helmsdale .... A mile or two further, and there comes in sight the old parish church of Kildonan. I was not prepared to find it the neat, trim, comfortable-looking little church it is. Once past the church, the township of Kildonan, the great centre of attraction, speedily

46. Artist's impression of the interior of Kildonan Parish Church which served as a refuge for many men down in their luck.

comes in sight, and a curious conglomeration of nondescript erections it looks. It consists of one street of about 150 yards in length, with a style of architecture more varied than elegant — wooden huts, canvas tents, inverted boats, old sails spread over the walls of turf, dilapidated vehicles covered with tarpaulin, old vans, strongly reminding one of travelling shows — anything and everything that the inventive genius or necessity of the owners could convert into partial shelter from wind and weather.

"My approach attracted the attention of one of the storekeepers who happened to be on the alert.

'Going to try your luck at the diggings, Sir? That's right. Plenty of gold here, if one could only get to it. I suppose you have no tent of your own, if you step in, and have a look at our crib, I think I can fix ye.'

"I stepped in as desired.

'There you are, Sir; this is our store. We shall be glad to supply you with anything you may want from a needle to a cradle.'

'A cradle!', I exclaimed in astonishment, 'What in the wide world do you want cradles for at Kildonan?'

47. "Rocking the Cradle" is a sketch by Mr John Campbell of Islay which he made during a visit to the diggings in 1869. It shows a section of a claim on the Kildonan Burn . . . "Waterworn drift arranged by running water in a groove carved in the edges of disturbed metamorphosed bent beds of Silurian rocks. Most of the gold is found near the rock amongst the biggest stones, and in chinks in the rock".

'Not for rocking babies, you may be sure. Meantime you had better step in at the next door and have a look at our cooking and sleeping departments.'

"The cooking and the sleeping departments turned out to be one and the same — a pretty sizeable room with an American stove at one end and on each side rude shelves that served for beds, built one above the other after the manner of berths on board ship — while a narrow deal table ran down the centre. I was informed if I had blankets I might have the use of one of these shelves — or rather the use of part of one of them — for 2s. per week . . .

"I laid in a sufficient supply of provisions and set out for 'the creek' to view the operations of the diggers. On reaching the summit of an adjacent hill, I could obtain a fine view of the burn for at least a mile, with the diggers hard at work. And a most picturesque and animated scene they presented — these hundred or so stalwart men, as they shifted to and fro — some digging with spades and shovels, some with picks pulling down the banks, some tearing up rock with crowbar, some carrying 'stuff' in buckets or wheeling it in barrows, some working sluices

48. Interior view of the store at Kildonan which provided accommodation for the miners at a rate of two shillings (10p) per week.

and 'long toms', some rocking the mysterious 'cradles' with the regularity of clockwork, while some with pick and gold pans were searching eagerly every nook and corner for 'prospects' of better 'claims'.

"Upon returning to the township I inaugurated my camp life by preparing my own tea. With scanty utensils for cooking — large common ware basins for teacups, and huge iron or pewter tablespoons instead of teaspoons — the bare deal in lieu of plates — no milk nor cream — rank, vile-smelling, hairy butter, and suet pretty much like the size and consistency of ordinary macadams — it may readily be imagined that my first meal at Kildonan was far from being a very enjoyable one. The diggers who lived in the hut — some twelve or thirteen in number — now dropped in one by one, and tired and hungry enough they looked.

"After supper there was a general adjournment, some to spinning yarns over their pipes, some to making necessary repairs, some to reviewing their day's work.

'Well, Ballarat, I suppose you have made a pile today?'

'O yes, Bill, I have, and a good one too — of rubbish.'

'How have you got on today, Bob?'

'Middling, just some five or six pennyweights.'

"Night creeps on, and there is a general scramble to bed, hard boards, each several inches apart, with common thin sacking filled with nothing by way of mattress, a blanket and transparent coverlet, both of limited dimensions and questionable cleanliness.

"No sleep for me, squeezed between two stout fellows who snored like porpoises, the frosty night air whistling in at every chink and cranny. Little wonder I could not sleep. Not many nights were destined to pass before I could sleep as soundly as the loudest snorer in the camp. About four o'clock I was startled by the information conveyed in language by no means of the choicest, that 'boiling' was ready, and that it was high time to rise. Then a scuffle for tea and coffee-pots, and the other varied utensils requisite for cooking and eating breakfast. Many were blessed with enormous appetites, and had plenty good substantial food. A few, with appetites probably equally keen, were under the necessity of contenting themselves with partaking but sparingly, and that of dry bread and coffee, oatmeal brose without milk or butter.

"Five o'clock sharp, and we are off, our pipes in full cloud, and half a loaf of bread and butter under our arm. My 'claim' is two miles away, but as the burn winds considerably, we can take a nearer cut across the hill. The morning is a misty and drizzly one, for it has rained hard during the night. It is very wet underfoot, but what of that when one is digging for gold! So off with coat and vest, roll up sleeves to the shoulder, for we must strip a 'paddock' sufficient in size to keep us cradling and washing for the day.

"Smoking? No time for that; I can't afford it; but a drink of water we must have and a glorious, cool refreshing draught it is. That? Why it is only a snake I have startled from this tuft of heather. Plenty of them hereabouts — and ugly, venomous creatures

43

they are — but we diggers know how to fix them — a cut with the shovel and all is over.

"Twelve o'clock! Dinner time already, and how little accomplished! A good three feet yet before we reach the bedrock, and that will take two hours at least, so we must have a short half-hour for dinner today. Doesn't the bread and butter taste delicious! and this crystal water — why it beats Bass's hollow! Seven o'clock, and our day's work is over, and not a bad day's work either, taking the ordinary run of luck. Over half a pennyweight of gold! About 1s 8d [8p] worth for fourteen hours' work! Nearly a penny and a half per hour!

"I have myself worked from five o'clock in the morning till seven or eight at night — worked up to the knees in mud and water, with hands blistered and hacked, and chipped and bruised, pulling down banks, tearing up rocks, rolling away great boulders, shovelling aside sand, gravel, and stones; and I have found in the end that I have earned about 9d or 10d [approx 4p].

"Lest, however, any of your readers might imagine that I have only been indulging in a fit of spleen owing to my own disappointment, permit me, in conclusion, to say, one having time and money at his disposal, but whose health may not be quite so vigorous as he would wish it, or who may desire to obtain a souvenir of Sutherland gold of his own digging, nothing could be more jollier or more pleasant than to provide himself with a tent and all the necessary concomitants, and to pitch his camp for a month or so during the summer at Kildonan or Suisgill Burns. . . . . I'll guarantee that he will soon pick up an appetite of the very first magnitude, which, after all is more to be desired than gold."

GOLD FROM KILDONAN
During the Kildonan Gold Rush of 1869, gold buyers established themselves in Helmsdale and throughout the year, gold changed hands at rates varying from £3 to £4 10s [£4.50] per ounce. Much of it went into jewellery; rings in a distinctive style known as "the Sutherland design" were made to special order by the jewellers of Inverness.

49. A pendant made from Kildonan gold and decorated by four garnets and a pearl from Brora.

In 1985, I learned of a small, gold cross made from Kildonan gold which was owned by a girl living in Edinburgh. Rona had received the item whilst she was a child from her grandfather who had taken part in the gold rush . . . but as an artist, rather than as a miner. It had always been in her family and its simple design enhanced its appearance. Artistically decorated with a Celtic motif, it carried four garnets clustered around a pearl from the Brora River. On the reverse was the single word "Kildonan".

Soon after, I tracked down a second cross. By prior appointment, I met the Bishop of Moray, Ross and Caithness in order to view a pectoral cross which belonged to the Countess of Sutherland, but which is on loan to the bishopric. This was also made from Kildonan gold and measured five inches in length. It was also decorated with a Celtic design which was featured around eleven pearls; for use, the cross carried a rope of green braid. It was a beautiful item and especially valuable in view of its origin.

In a pleasant interview with the bishop, he confirmed that he wore it on official engagements outside Scotland.

More recently, I have directed enquiries to locating a watch which is also made from Kildonan gold. It was presented to Robert Nelson Gilchrist by the Duke of Sutherland in 1869 to acknowledge Gilchrist as the man who had discovered the gold the previous year. The presentation ceremony was well reported in the newspapers at the time and we have an accurate description of the event:

50. The Bishop's cross.

The Duke of Sutherland paid a visit to Kildonan (2 June 1869) when he took the opportunity of presenting to Mr Robert Gilchrist, as the discoverer of the gold, a valuable gold watch, which Mr Gilchrist will, no doubt, prize very highly as a memento of the Duke's kindness and generosity. The watch was inscribed — "Robert Nelson Gilchrist the discoverer of Sutherland Gold, from the Duke of Sutherland. Nov 1868."

His Grace was accompanied by Sir Roderick Murchison; the Rev Mr Joass, Golspie; and other gentlemen.

From time to time, newspapers report "sightings" of this watch and personal enquiries have taken me the length and breadth of the United Kingdom — from Kildonan to Aberdeen, from Fraserburgh to Birmingham, from Glasgow to Scarborough. The trail finally went cold in Cambridge and it is possible we will never find out if the famous watch has survived its hundred years of history.

The current prospectors on the Kildonan Burn are relatively

modest in their demands and most members of the Kildonan Prospecting Club have made sure of a souvenir of the long hours of hard, physical effort, by converting about four grams of gold into a ring for themselves and their partners. One exception is the owl which Alf Henderson commissioned for his wife from gold which he had collected over a number of summers. There is a good chance the owl will survive the next hundred years as Alf has already bequeathed his collection of native gold to the British Museum in London.

NUGGETS

It is not easy to define a nugget.

One dictionary states that a nugget is a "lump of gold" and suggests the word may have originated from the Swedish 'nug' meaning lump or block. In Australia 'lumps' of gold weighing up to 200 lbs have been found and the Californian Gold Rush was based on the search for nuggets — and, at the time, little else.

It is no surprise that the Gold Information Centre in Madison Square, New York, provides a good explanation:

> *Gold Nugget:* A water-worn mass of placer gold (a form of natural gold) washed from the rock that contained it and deposited in riverbeds. Usually ranging in weight from approximately 30 grams to 50 kilograms. The heaviest nugget ever recorded, named 'The Welcome Stranger', was found in Australia in 1869 and weighed 90.9 kilograms.

In the British rivers, nuggets are few and far between . . . and considerably smaller in weight. Good criteria for assessing a nugget . . . can it be heard rattling in a bottle? Can it be picked up between finger and thumb?

One hears more about nuggets than actually seeing them. This may account for the amiable convention of assigning names to some of the more illustrious ones. If nothing else, it helps to identify a specimen in a convenient manner and the practice was revived during the Kildonan Gold Rush of 1869.

Writing in *The Inverness Courier* for 22nd April 1869, a reporter describes The Rutherford Nugget:

> Dr Rutherford of Helmsdale yesterday presented to the Duke of

Sutherland the largest nugget yet which has been found at the Kildonan Diggings. It was got some time ago by quite a young lad from whom it was bought by Dr Rutherford for the sum of £9. The nugget is singularly free from earthy matter, and weighs two ounces 19 grains. It was exhibited for a while in the shop of Messrs Ferguson Brothers, Inverness, at which instance it was photographed by Mr D Whyte, Inverness.

The Rutherford Nugget is also known as "The Nine Pound Nugget" and a drawing of it is reproduced in a pamphlet entitled "Something from 'The Diggins' in Sutherland" by Mr John Campbell of Islay.

Other Sutherland nuggets have earned themselves proper names and an illustration.

"The Helmsdale Nugget" is on display in the Mineral Gallery of the British Museum as specimen BM34951. It was found by a

THREE SUTHERLAND NUGGETS

51. The Rutherford Nugget which is sometimes known as 'The £9 Nugget' on account of the initial purchase price. By September of 1869, duplicates were being sold in London for 2s 6d (12½p) by a mineral dealer.

52. The Helmsdale Nugget was found by a shoemaker from Elgin by the name of Scott and he sold it to a Mr Kennedy who sold it, through an agent James Mackay of Elgin to the British Museum in 1870. It weighs about 40.5 gm.

53. The Sutherland Nugget is a model of one in the possession of the Duke of Sutherland which was acquired by the British Museum from Professor J Tennant in 1879.

51

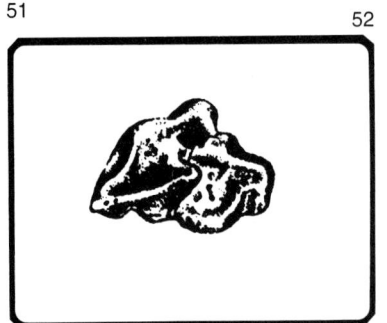

52

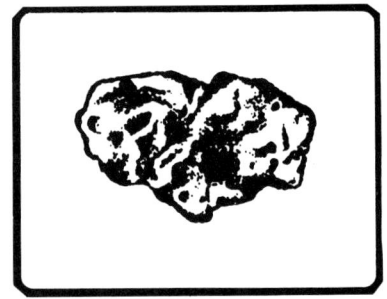

53

shoemaker from Elgin by the name of Scott, and he sold it to Mr Kennedy who sold it, through an agent James Mackay of Elgin, to the Museum in 1870. It weighs about 40.5 gm.

"The Sutherland Nugget" is in the possession of the Duke of Sutherland but a model was acquired by the British Museum as specimen BM87549 from Professor J Tennant in 1879. It is possible that this and the Rutherford Nugget are one and the same.

One of the strangest British nuggets is "The Gemmel" which was found at Crawfordmoor by a Leadhills man in 1872 but continued to excite interest and speculation for more than six years. It was frequently cited as the proof that there was gold-bearing quartz in the locality of the Wanlockhead and Leadhills villages. Of course, this claim had its critics who were not averse to suggesting that it had been 'imported' into the district — with a view to "palming it off to a gentleman".

When Patrick Dudgeon prepared "Historical Notes on the Occurence of Gold in the South of Scotland" in 1875 in order to raise funds at a Glasgow bazaar in aid of Miss Clugston's Home for Incurables,

54. The cover for a pamphlet published in 1875 in order to raise funds at a Glasgow bazaar for Miss Clugston's Home for Incurables which shows the Gemmel Nugget found at Wanlockhead in 1872.
The man unfortunately broke the specimen into a number of pieces, and sold them to various individuals in the district. With the assistance of Mr Clark of Speddoch, I obtained the loan of all the pieces from their respective owners and joined them together; only a few chips were found wanting, and the drawing which is the exact size of the specimen was made from the restored mass.
Mr Patrick Dudgeon, 1875.

THE GEMMEL NUGGET

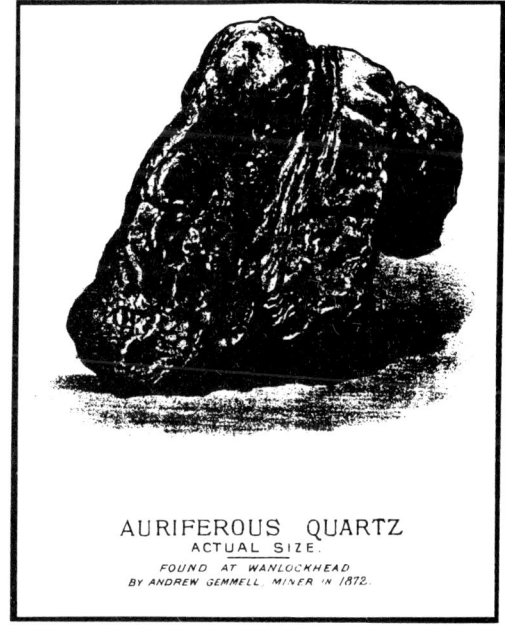

AURIFEROUS QUARTZ
ACTUAL SIZE
FOUND AT WANLOCKHEAD
BY ANDREW GEMMELL, MINER IN 1872.

49

he included a coloured lithograph of the nugget as a frontispiece. In the pamphlet, he explains the circumstances of the drawing:

> The very interesting specimen of auriferous quartz — for the illustration of which I am indebted to the kindness of His Grace the Duke of Buccleuch — was found by a miner at Wanlockhead in 1872. The man unfortunately broke the specimen into a number of pieces, and sold them to various individuals in the district.
>
> With the assistance of Mr Clark of Speddoch, I obtained the loan of all the pieces from their respective owners and joined them together; only a few chips were found wanting, and the drawing which is the exact size of the specimen was made from the restored mass.

PART III
APPENDIX

APPENDIX

PANNING

The basic method of gold recovery is by 'panning' — and the techniques used today are little different from those used in the great gold occasions of history . . . from the time of the Pharaohs, and the Aztecs, to the more recent gold rush of California in 1849. Early pans were made of wood; some were conical in shape and have the name 'batea'. In the 19th century — and especially in California, Scotland, Australia and South Africa, — pans were fashioned from metal. Nowadays, plastic is being used to provide the basic requirements in the design as well as other features to make the washing stage more enjoyable — and in view of the paucity of gold in Great Britain, more reliable!

As a general rule, all washing follows the same routine. The pan is loaded with gold-bearing gravel, and using plentiful supplies of water, the contents are agitated so as to mix them thoroughly. During the preliminary phase, it is reasoned that particles of gold will slowly move through the rocks, the pebbles, the clay, the sand and water

55. An often neglected aspect of panning is to find a comfortable seat with easy access to slow moving water of adequate depth.

56. Chris Engels of Aberdeen is a panner of repute who eschews the modern plastics in preference for a traditional metal pan. A portrait taken on the Afon Wen, 1983.

to the bottom of the pan. It is not unusual to carry out this stage by kneading the contents of the pan with both hands so as to break up lumps of earth and to free gold from any clay-like material. As the large lumps disintegrate, it is appropriate to discard large stones and pebbles so as to reduce the weight of the pan's contents.

When the material in the pan has been thoroughly broken up and all the large rocks have been removed, take the pan in both hands and flood it with water. With the hands positioned slightly to the back of the middle of the pan, find a comfortable position which supports its weight but ensures — for manipulation — that the thumbs are grasping the top edge.

At this point the washing begins. Give the pan a slightly oscillating, circular motion and gently shake the contents from side to side. At the same time, tilt the pan downwards so that the lighter weight material will wash away. In the early stages, the lighter material can also be scraped away and removed over the edge of the pan — preferably with the heel of the hand.

This stage is repeated over and over again until only the heavier materials (that is, black sand and gold) remain in the pan. If this can be reduced further by careful panning, so much the better. If gold

57. Patrick Reeson of North Wales panning the gravels of the Kildonan Burn, June 1987.

can be seen with the naked eye, some people like to transfer it to a glass phial of water. A camel-hair brush is a help in doing so, but 'experts' will achieve the same result by drying a finger and lifting the gold by adhesion. An alternative to this procedure is to transfer the 'concentrate' (that is, the black sand AND the gold) to a bucket so as to press on with the gold recovery. At a later date, and usually at home, this concentrate is carefully panned out so as to recover the gold under ideal conditions.

It is worth bearing in mind that the panning stage might take from ten to fifteen minutes and the consequence of impatience is usually to lose some gold. The 'old timers' would wash ten pans each hour during a ten hour day. Another oft-quoted statistic is that a cubic yard of gravel will take about 150 pans to wash.

There are three rules of panning which help to prevent the gold miner from going crackers:

* Trust the pan
* Use plenty of water at ALL times
* Visualise the passage of a particle of gold during the washing stages. Gold is so heavy it will prefer to travel to the base of the pan rather than 'jump' over the sides, as so many beginners expect.

And finally . . .

Just once in a while, re-wash your 'tailings'. These are the accumulations of gravel which should be developing on the river bed close by your site of working. If you find nothing, no harm is done; if you find some gold, it's a bonus!

THE HENDERSON PUMP

In recent years, the gold panning fraternity has experienced a remarkable innovation — a new piece of equipment has been introduced into what is a very tradition-based activity. This is "the pump". Alternative descriptions would include the sucker, the syringe, or the syphon — but throughout Europe, its fame has spread during the past five years and it has become known as the "pump".

Less well known is that Mr George Alfred Henderson produced the first prototype in 1985 and readily acknowledges that he modified an idea from a device first seen on the Isle of Man. Since then, not only has Henderson manufactured up to twenty other models, he has refined the design to a very valuable accessory for the British gold prospector. [See Note 1] In addition, numerous people have built their own versions; very often enterprising improvisations are used to achieve Henderson's working principle.

The "case" is a tube of plastic drain-pipe available from DIY stores, to which is added a "nose cone". Initially, this was adapted from a spare part for an industrial vacuum cleaner, but nowadays supplies of this component are in short supply. Running through the centre of the tube is an effective washer system which is fastened to a rod controlled by a handle from the rear. There is a temptation to make the tube too long and although prototypes measuring four feet long have proved ideal for accessing sites in deep water, these present a problem in operating the handle effectively. If the tube is too short, some of the benefit in accessing out-of-the-way places is lost. The optimum length is around 30 inches.

In selecting a length it is a wise precaution to take into consideration how it will be carried on overseas trips and to tailor the length to fit a medium-sized suitcase. The author has taken advantage of this aspect on visits to France, Switzerland, Austria, Lapland and America, — and on occasions where space is at a premium, the central tube of the pump has been stuffed with socks and underwear.

The washer arrangement consists of two circular pieces of milled plastic which sandwich a leather washer. The latter overlaps one of the plastic washers and ought to be renewed every few years with a fresh piece of leather.

When Henderson first produced his pump, he provided an aperture on the nose cone of about 1.5 inches diameter. In time, with extensive use in gravel and sand, this widened and the designer then realised that, with a good vacuum seal at the washer, it was permissible to increase the size of the opening to advantage. This is, of course, only possible where the vacuum pressure is reliable, but it does much to reduce blockages by awkwardly shaped stones.

Operating the pump is simple and hardly needs further comment. It should be used methodically so as to identify gold-bearing "hot spots". The water-gravel-gold mixture is best brought into the tube by a quick action, but the handle should be used sparingly during the discharge stage. By adopting a rhythmic routine, and alternating the pumping and panning sessions, it is possible to process a vast quantity of river gravel without too many muscular ill-effects.

In developing the principle of the pump and generously helping to make the design freely available, Alf Henderson has done much for the recreation of gold panning.

Notes:
1 The reason why the pump is especially suited to British rivers is that the sluice box and/or the dredge are not popular with many landowners and, thanks to the efforts of the Kildonan Prospecting Club in the mid 1980s, the pump is now accepted as being "environmentally friendly".

2 It is readily accepted that for many years the principle of the pump has been used to retrieve gold from cracks in rocks by using, for example, rubber syringes intended for topping up the cells of batteries, and that American gold prospectors use a "sniffer". Bad designs generally meant that the forward compartment had to be removed in order to access the gold.

PROSPECTING WITH A SLUICE BOX

To graduate from goldpanning, one turns to sluicing with a sluice box. This is a simple, portable piece of kit which is "fitted" into the river, and when properly set up, it takes much of the strain out of gold recovery. It is claimed that a sluice box will process five to ten times the amount of ground that can be done with a pan in the same time. Of course, designs vary and will range from the heavy, permanent installations used by professionals in 19th century California and in 20th century Lapland, to the lightweight models made of aluminium, or light plastic, which are easily transportable, easily set up, and — when the time is right, easily cleaned up.

The aim of sluicing is to catch and make use of the natural flow of water by positioning the sluice box in a stream relatively close to the work place. The preliminary stage generally takes some time and it is not unusual to use rocks to support the sluice, or to hold it in place with a large rock on top. The angle of water-flow is adjusted so that small rocks and stones are transported along the sluice box by the force of the water. It may be necessary to construct a "wing dam" close to the sluice in order to direct more or less water as required. Gold is also carried along in the same way by the force of the water but a series of riffles embedded in the base is intended to ensure that the precious metal is trapped securely until the end of the session.

Once set up, the gold miner works his sluice box methodically; charging the head of his sluice with auriferous gravel and concentrating on *helping* the water to progress the material through the run of the sluice. From time to time it will be necessary to clear "tailings" from the end of the box but, in general, gold recovery is reserved for the last stage in one grand panning session. The "concentrates" which have accumulated in the riffles of the sluice box are carefully transferred to a pan and washed in the normal way.

For the technician, the design of the sluice box presents an absorbing challenge and many different models exist in which a special "throat" is used to control the ingress of water, and special channels contain riffles of different design so as to differentiate between fine and coarse gold.

In spite of an increase in gold panning in the United Kingdom, only a few specialists are working the rivers with sluice boxes and this is, in part, due to the restrictions placed on "mechanical devices" on some of the properties. For example, the Suisgill Estate in the Strath of Kildonan has a total ban on their use in the Kildonan and Suisgill Burns. The Wemyss Estates, which manage Glengaber on Henderland Moor, have announced the intention of restricting permission to gold panning only. At certain times of the year the sluice box should not be used on the gold bearing rivers of North Wales in view of the possibility of the increased amounts of debris which could affect the salmon breeding.

Fortunately, the rivers of Crawfordmoor — Elvan Water, Mennock Water, for example, — have no restrictions and, appropriately, this freedom allows the gold prospector to continue a tradition which was established by the gold-seeking lead miners in the early part of the twentieth century. In their spare time, and especially during the General Strike of 1926 when the lead mines were forced to close, the miners would turn to gold and favoured a "trough" for the recovery of gold from the local rivers. This was a simple form of sluice box which had an adjustable gate to control the water through the box. A paddle was used to move the gold-bearing gravel up to the head of water — whereupon the gravel was sorted into the expected component parts and was either washed away, (sand, stones and pebbles) or was trapped in a groove (black sand and gold).

THE ELECTRONIC PROSPECTOR

In the age of the microchip it is not unreasonable to look for some help in the field of gold prospecting — and a commonly asked question poses the possibility of finding gold with a conventional metal detector. Whilst it is a reasonable question to ask, it is a difficult task to set the instrument.

In Britain, gold is generally distributed in such minute quantities that it is transparent to the types of detector which are designed and equipped for locating specimens such as coins, old bullets and other treasure trove. Only in the United States of America do advertisements make claims that the metal detector will identify large nuggets.

Nevertheless, modern technology has managed to contribute a very sophisticated aid to locating gold — the electronic earth probe. Gold is no ordinary metal and a Swedish inventor, Lars Guldstrom, has developed a theoretical principle which he has turned into a practical reality. Guldstrom's invention is known as The Goldspear and it works by selectively reading the electro-chemical potential and conductivity of minerals buried in the earth. A novel aspect of his circuitry design is an electronic programme which provides a remarkable sensitivity just where it is needed — in detecting single fragments of gold down to a size of 300 mesh. (That is, 0.002".) It will separately identify 'black sand', a material universally accepted as a good indicator for gold to be in the vicinity. In auriferously rich areas, the signals indicating "general minerals" can be switched off — in order to pinpoint the two specific minerals of gold and platinum.

So as to differentiate between different materials, four signals are provided by the instrument. Coloured light-emitting diodes provide a visual cue and range from green through red to orange and finally to yellow in the presence of gold. The lights are also reinforced by different audible signals — but with experience, one relies on the latter for the qualitative assessment of a prospect and only refers to the coloured lights for confirmation. Thus, black sand is associated with a crackling noise whereas gold qualifies for a distinctive 'bleep'. If necessary, these audible signals can be tuned out and routed to the user through a set of headphones so as to eliminate other distracting sounds such as wind, traffic and rushing water.

The Goldspear is made up of two complementary units. The nerve centre is a compact box packed with electronics to which is attached a four foot metal probe carrying the detection unit at its tip. In order to access out of the way places, extensions can be added to the probe which make it possible to research a river bed 20 feet deep.

It is very simple to operate. At a potentially auriferous location, the probe is plunged into the ground. In the presence of gold, heavy minerals and/or black sand, there will be signals identified by the electronic unit. By adjusting the controls, the signals from black sand and heavy minerals can be tuned out so as to make an assessment of the gold deposits only. The real advantage of The Goldspear materialises when it is used methodically for 'prospecting'. That is, the probing is carried out over a carefully selected area and a "map" (which can either be real or just stored in one's head) is built up from which the most fruitful spots for panning can be determined.

Another benefit of the The Goldspear over traditional prospecting is the ease with which it can determine the depth of a layer of gold deposit. In many rivers, (for example, the River Mawdach in Wales, the Hangas River in Northern Finland and the South Fork of the American River in California), the most available gold is not on bedrock but often exists as a deposited layer . . . possibly one to three feet below the surface of the gravel bed. Without prior experience, it is possible to chance on this layer and accidentally work through it — whereas analysis with the The Goldspear would indicate the merit of working laterally in such circumstances.

Of course, some people will argue that The Goldspear takes much of the fun out of recreational gold panning but, on the other hand, no-one can deny that its availability has harnessed modern technology into a novel opportunity for the enthusiast.

Note

An often heard criticism of the The Goldspear is that it is asking a lot of the probe to provide a true assessment of gold deposits in view of the requirement that the tip must make contact with gold for a signal to be issued. This is to assume that there is a ration on the permissible number of times the probe can be inserted in the ground at any one spot. In practice, when using the outfit, one very quickly develops a technique of probing the ground repeatedly and, by instinct, averaging out the nature of the individual signals.

RIVERS AND RIVER LORE

All books on gold recovery include advice about the river, the course of the river, and the evolution of the river. Having this knowledge is fundamental to the selection of potential spots for gold panning. After all, it is one thing to know that a river is auriferous but quite another matter knowing just where the gold may be found in the river.

Experts try to identify the places where gold will accumulate, (and these are called the "placers",) so as to eliminate the unlikely locations from their assessments. To understand the origin of gold involves starting with some geology and accepting that three to ten million years ago, the material which nowadays forms gold would be forced upwards from the earth's interior into fissures and cracks in the bed rock.

During the cooling process, the molten material, known as "magma", would separate into veins of gold and the milky-white rock material we call quartz. In this form, gold is described as "lode ore" and is recovered by mining, which is followed by the attendant processing techniques of milling, crushing, separating and refining. When the same work is done by nature, placer gold is the end result and sooner or later, the gold finds its way into the nearby rivers.

It is not difficult to visualise the next sequence of events. When the heavy spring floods create spate conditions in the rivers, the water will have scoured gold from the hillside along with rocks and other debris. Once in the river, a long process of fragmentation begins as the quartz rolls, bounces, tumbles and impacts on other rocks. Most of the time, gold is being broken up but it may experience an alternative change by being pounded into chunks which are known as "nuggets".

While the water is running fast, gold and the heavy black sand will continue to move along with it. Whenever the current slows down and the pressure eases, such as, at bends in the river, or where the gold meets obstacles, such as huge boulders firmly stuck in the river, the gold will drop out and settle in gentle water. Gold is heavy and will follow the easiest route downstream but will stop wherever convenient.

Even after the gold drops out of the current and settles at the bottom of the river, the movement of the water is adequate to provide the energy for the gold to continue to sift through the "overburden" until it comes to rest on the bedrock. Very often gold will be carried further along the bedrock until it eventually becomes trapped in a crack or crevice with some permanence.

The above reasoning applies conveniently to most of the rivers in Britain, such as the Avon, Wen and River Mawdach in North Wales, the famous rivers of Crawfordmoor, (Mennock Water, Elvan Water and the Meggat Water,) the Cononish of Central Scotland and others.

In the case of the Kildonan Burn and the Suisgill Burn in Sutherland, however, it is more likely that the gold originated over a land mass which is now Iceland and was transported by glacier. The grinding processes within the movement of the glacier would produce a fine gold which is deposited along with the alluvium to give us the alluvial gold of today. In appearance, there is no difference between "alluvial gold" and "placer gold".

ALL IS NOT GOLD . . .

The story of gold discovery in Britain would be incomplete if no reference was made to two important strikes which caught the public imagination at the time, but failed to materialise as anything other than a false alarm.

The first had its origins in a penal colony in Australia. In 1852, a convict who was working in a gold mine, wrote to friends in Fife,

Scotland, and advised them that he had often seen material at home in the lime quarries above Kinnesswood in the Bishop's Hill, which was similar to the material he was digging in Australia. At this particular time, the Californian gold rush was at its height and everyone dreamed of gold on their doorstep. No wonder that the prompt from Australia sparked off a minor gold rush in Scotland. Contemporary reports speak of "the hill dotted with hundreds eager to try their hand at digging" and because of the number of tents erected on the hillside, it presented the appearance of "an immense public fair being held".

There was a daily average of three hundred diggers and many of them were local coal miners and weavers who had abandoned their trades "to embark on the alluring lottery of gold seeking". The good feature of the incident is that the adventurers came to look for a gold-like substance and on finding "a bed of ochre, abounding in globular masses of iron pyrites, known to the quarrymen as 'fairy balls', from the size of a fist to that of a man's head", they were more than satisfied with their efforts.

The other astonishing discovery took place in the suburbs of Edinburgh whilst an extension was being built to Leith Hospital in 1901. As the workmen dug the foundations, they came on beds of clay which carried "gold bearing quartz". The Edinburgh newspapers of the day enjoyed reporting the bonanza on their doorstep; traffic was disrupted and the authorities moved in to consult many experts. A gold mining manager, home from the Transvaal on holiday, came to see the quartz and "having his prospector's glass with him" was asked for an opinion.

The Mining Journal reports the event:

> He was confronted with a very good sample of the quartz unearthed from King Street and he gave it his undivided attention for twenty minutes without uttering a word. Then slowly he raised his head and said, "I'm not prepared to say it isn't gold, but I don't think it is."

He admitted the quartz bore a strong resemblance to that found in South Africa, but he did not think it was gold. "If it is gold", he concluded, "it is very rich indeed." A sample was sent for assay to Glasgow, but the directors of the hospital, expressing a reluctance to open a gold mine, declared their intention to proceed with their building work and gave orders to "pour the cement". To this day we are left with a conundrum and a legend that Leith Hospital is built on a gold mine!

References
1. The Fifeshire Gold Diggings of 1852 Edinburgh Geological Society 1869 W Lauder Lindsay
2. The Discovery of Gold in Scotland *The Mining Journal* April 1901

THE FUTURE

It is fair to say that gold prospecting has been practised in Britain for over two thousand years but, regrettably, the yellow metal is a precious material and a natural resource which is not being replenished. It could be hoped that some benefactor might grind up ten kilograms of gold bullion and seed all the rivers on the fourth degree with generous handfuls. In this way gold washing could see a renaissance and might last a further two thousand years. But, in reality, it is a dream and not only is gold becoming more difficult to win than, say ten years ago, as the recreation becomes more popular, more restrictions are being placed on access to rivers, to land and to the river banks.

Already Suisgill Burn in Scotland is a Site of Special Scientific Interest and rumours suggest that access to the Kildonan Burn will be restricted in the near future in order that commercial mining can be carried out. At Dolgellau, environmental restrictions imposed by the Forestry Commission and the Welsh Water Authority have always been present but recent years have seen a tightening up of those rules which help to preserve the ecology.

Most of the historical rivers of Crawfordmoor are on privately owned property with public access, but the landowners are becoming increasingly aware of the numbers of legitimate goldpanners as well as an increase in the numbers who utilise mechanical equipment, such as dredgers imported from the USA. For example, and with some justification, curbs are being placed on the enthusiasm of those people who believe that it is acceptable to pan the waters of Glengaber Burn on the grounds that "there is nothing to prevent you doing so."

The unique and precious gold vein at Hope's Nose in Devon has been a Site of Special Scientific Interest for many years and the Roman gold mines at Dolacothi are now in the possession of the National Trust. Even although it has no reputation as a gold panning location, it is not likely to start now that it belongs to the nation.

On the other hand, the very popularity of gold panning and its ancillary interests is resulting in new initiatives which have been available for many years in countries which do not have a tradition of gold discovery. At theme parks in Denmark, for example, it is possible to receive tuition in panning and to practice the recovery of a "gold dust" which is probably a form of iron pyrites. The panned sample is intended to be returned to the organiser who "processes" it in complex-looking machinery which produces a token which reminds the visitor of his visit to the park.

On special occasions, Wanlockhead Museum Trust has provided opportunities for children to pan in the waters of Wanlock Water. The "gold" is fool's gold which has been collected from the nearby lead mines, but the circumstances and surroundings are authentic and help to reinforce a tradition of the locality and to complement the displays in the Museum of Scottish Leadmining nearby. More recently, the Clogau Gold Mine Centre in North Wales has introduced panning as an attraction for children and an innovative feature of the provided paydirt is the selection of semi-precious stones which are included — and which are retained by the panner.

In Cumbria, The Lakeland Mines and Quarries Trust in Threlkeld, Keswick, strikes a more serious note by its arrangements for panning. An instructor is on hand to provide advice and tuition, and the bag of gravel is guaranteed to contain real gold so that the experience in handling the pan is authentic and genuine. There is no reason why the idea of gold panning attractions at theme parks and museum trusts should not be developed further. It has the advantage that it is conducted out of doors and that the necessary facilities are not too demanding on the organisers. One can predict a growth in venues and the stylised surroundings are not dis-similar to those in force at world goldpanning championships. That is, competitors are allocated a restricted space, and are provided with a quantity of gravel which has been deliberately seeded with grains of gold by the jury of the organisers.

Each year the World Championships go from strength to strength and some of the most enthusiastic and successful competitors are those who do not have the opportunity to pan in rivers but have to rely on gaining experience at clubs, in tanks, in ponds — and in some cases, in the bath at home! It is a fair conclusion to say that the outlook for sustained gold panning could be more favourable at this point in the twentieth century, but it could be a lot worse!